# Contents

## GROUP A: GLOBAL INNOVATORS OF INDIAN ORIGIN (2024)

## GROUP B: AGRICULTUTE (2024)

# TOP 100 INDIAN INNOVATIONS 2024

## INNOVATION YEARBOOK SERIES

Indian Innovators
Association

## GROUP C: HEALTHCARE (2024)

## GROUP D: WATER & ENVIRONMENT (2024)

## GROUP F: INDUSTRIAL SOLUTIONS (2024)

## GROUP G: CONSUMER PRODUCTS



## YOUNG INNOVATORS (2024)

# Acknowledgement

This is the third publication of the Innovation Yearbook series. The first edition- Top 100 Indian Innovations (2022) provided 100 answers to the question- Is India innovative? And with the next 100 profiled in 2023, the Innovation Yearbook has been accepted as the most authentic reference volume on Indian Innovations. With the current volume, the count goes to 300 Indian innovations. In addition to serving as a refrence book, we hope the innovation yearbook series will be widely distributed in the libraries, to inform & excite the students. Each volume explains technology trends, patent status, innovation merits, and the personal background of each innovator. This type of comprehensive information is not available to seekers in any textbook or social media

We received nearly 1000 nominations, directly or indirectly, and thank all innovators & stakeholders for helping us in this task. A hundred innovations are selected from out-of-the-sourced documents. For selection, equal weight is given to 'innovator profile' and 'innovation merit'.

To supplement information received along with recommendations/ nominations, additional details are collected from publicly available documents on the net. Photos of inventors/ founders are taken from their LinkedIn profiles which are also cited in select cases. The photo on the cover page is provided by Zen Technologies.

**Editors**
Aynampudi.Subbarao.Rao,
KVSP Rao,
Sachinn Aggarwal
Raman Teja Venigalla

# Foreword

I am delighted to introduce the third edition of the *Innovation Yearbook Series: Top 100 Indian Innovations (2024)*, a compendium that celebrates the remarkable ingenuity and creativity of India's brightest minds. This year's edition, like its predecessors in 2022 and 2023, showcases the groundbreaking work being done across a diverse range of fields, from quantum computing and sustainable energy solutions to biotechnology and advanced materials.

In today's rapidly evolving world, innovation is not just a buzzword; it is the lifeblood of progress and sustainability. The solutions documented in this yearbook represent the cutting edge of technological and scientific advancement, reflecting India's increasing prominence on the global innovation stage. The inventors and entrepreneurs featured here are not only pushing the boundaries of what is possible but are also addressing some of the most pressing challenges faced by society today—be it in healthcare, agriculture, energy, or digital transformation.

It is particularly inspiring to see the diversity of thought and the spirit of collaboration that permeates these pages. The contributions range from young startups challenging the status quo to established organizations leading the way with pioneering research and development. This mix of ideas and approaches is vital, as it fosters a robust ecosystem where innovation can thrive.

At IBS Global, we firmly believe that fostering innovation is key to driving sustainable growth and development. We are proud to support initiatives that encourage creativity and entrepreneurial spirit, such as this yearbook sponsored by the Indian Innovators Association. By providing a platform for innovators to showcase their work, we hope to inspire others to think boldly, act courageously, and continue pushing the frontiers of knowledge.

As you turn these pages, I encourage you to reflect on the incredible potential that lies within each of these innovations. They are not just ideas but solutions that have the power to transform industries, improve lives, and shape the future of our world. It is our hope that this collection serves as a source of inspiration, knowledge, and motivation for all who seek to make a difference.

Congratulations to all the innovators featured in this edition. Your passion, dedication, and relentless pursuit of excellence are truly commendable. We look forward to seeing how your innovations will continue to evolve and impact the world positively.

Here's to a future driven by innovation, collaboration, and progress!

Sincerely,
**Ms. Edyta Wolczyk**
CEO, IBS Global,
Poland

# Preface

Innovation is driven by demand. Though India is late adoptor of the Demandside Innovation policy, the results started showing.

The function of innovation is threefold. The first function of innovation is to drive economic development, which in terms of policy and analysis is still largely associated with the nation-state. It is critical to note that the economic dynamics of countries depend as much on demand—that is, on the speed of adopting and absorbing innovations—as they do on the generation of innovation itself. In fact, for a considerable time economists have regarded favourable conditions for innovation diffusion as the most important driver for economic development. The constructive role of lead users in testing, further improving, or even co-generating innovations is an essential element of these favourable demand conditions. Thus systems with an advanced demand for innovation offer better context conditions for firms to invest in innovation, often leading to export advantages as international demand catches up.

Second, innovation systems need to help satisfy national and local needs. In market economies, needs are fulfilled only if they are articulated as demand—that is, as signals to potential suppliers to buy for a certain price. Innovation systems are of limited legitimacy if the innovation they offer cannot respond to the needs of their own populations, that is, if they are not orientated towards local demand.

Expanding this understanding of innovation as serving needs on a global scale leads us to the third function of innovation: Innovations are essential for tackling big global challenges. However, simply producing ever-more sophisticated technologies that are not rolled out broadly and globally will not be sufficient to tackle global challenges such as the reduction of carbon emissions. For that to happen, broad diffusion and application of innovative energy-efficient products and processes are required. This means that demand must be articulated and must connect with supply, and potential buyers and users must be able to understand and use innovations that address the challenges defined.

Four major ways in which demand influences innovation:

1. Changes in demand may trigger innovation in traditional way (demand pull),

2. Demand may be responsive to innovations offered by the market place (supply push),

3. Users and producers may co-produce innovations,

4. Users produce innovations themselves, for their own purposes, but with a potential to spread across markets.

Demand-side innovation policy instruments may be defined as a set of public measures to increase the demand for innovations, to improve the conditions for the uptake of innovations, and/or to improve the articulation of demand in order to spur innovations and the diffusion of innovations. Such a broad definition of demand-based innovation policies implies twin rationales, namely, to promote and stimulate innovation and to increase the diffusion of innovation. In addition, this second rationale, the diffusion of innovation, further implies that the concept of innovation extends beyond the scope of 'new to the world' and encompasses innovations that are 'new to a firm' or to a certain geographical space.

The vast majority of scholars working in science, technology, and innovation (STI) policy in developing countries agree that traditional supply-side STI policy has failed to deliver economic development, and in particular has failed to include the poor. The demand-side instruments available for the state are numerous, but they can be classified into five groups where strategic demand-side approaches can combine those measures and ensure that corresponding supply-side measures are in place:

➤ The state can act as buyer. The most direct leverage for the state is public procurement of innovation, whereby the state strategically decides to invest in innovations that help to satisfy societal demands or make public services more effective and efficient.

- ➤ The state action can apply so-called price-based measures. Subsidies or tax allowances reduce the price for innovations in their early stage in order to set in motion a virtuous cycle of diffusion and cost reduction through economies of scale.

- ➤ There are numerous non-financial measures by which the state can improve the capabilities and readiness of potential customers to buy and use an innovation. Those instruments include awareness measures, labels and demonstration projects to build up trust in innovations, and education programmes designed to enable consumers and firms to use innovations effectively.

- ➤ The state can support the articulation of needs (e.g., through needs-based foresight activities); translating those needs into signals of demands for innovation is important to direct innovation activities towards demands.

- ➤ The state can support the user of innovation in generating or cogenerating innovation, including so-called social innovation initiatives.

We illustred 300 innovations in our innovation yearbook series. Readers will notice the common thread - demand for innovations created or enabled by the state policies.

A.S.Rao
Indian Innovators Association

# Global Innovators of Indian Origin (2024)

# Agzen.com-precision spraying-Kripa Varanasi, USA

Kripa K. Varanasi is a Professor of Mechanical Engineering at MIT, where he earned his M.E. and Ph.D. after completing a B.Tech at IIT Madras. He has co-founded several companies, including LiquiGlide, Dropwise, Infinite Cooling, and Veron24. LiquiGlide has been recognized by Time and Forbes magazines as one of the "Best Inventions of the Year." (https://meche.mit.edu/people/faculty/kripa@mit.edu)

## Technology

For about 80 years, farmers have relied on general practices for spraying agricultural chemicals, often based on past experiences. To improve deposition efficiency, several strategies have been developed. One involves upgrading the spraying apparatus, such as using electrostatic spraying and reducing pesticide droplet size. Another strategy is to enhance pesticide formulations with polymer additives and surfactants. A third approach focuses on altering the target surface properties in-situ during spraying by creating sparse hydrophilic defects on the substrate. A final strategy aims to improve droplet deposition through nanofiber networks that prevent splashing and bouncing.

## Innovation

upported by the MIT Tata Center and the Abdul Latif Jameel Water and Food Systems Lab, Varanasi and his team studied how droplets interact with plant surfaces to enhance application efficiency. This research led to the development of a nozzle system that coats droplets with compounds, increasing their retention on leaves. The product, called Enhance Coverage, applies in real time as the sprayer moves through fields.

To market this system effectively, AgZen needed to accurately measure the amount of spray adhering to plants in real-time. They developed a solution tested on farms that involves a unit attachable to any sprayer arm. This unit features two sensor stacks, positioned before and after the sprayer nozzles, with software that provides real-time data on spray coverage. It calculates how much the droplets will spread or evaporate, giving a precise estimate of final coverage.

## Patent

Title: Systems and Methods for Real-Time Measurement and Control of Sprayed Liquid Coverage on Plant Surfaces

Patent Number: 11921493

Inventors: Vishnu Jayaprakash, Kripa Kiran Varanasi

## Commercialization

Prof. Kripa Varanasi and Vishnu Jayaprakash co-founded AgZen, where Varanasi serves as Chairman and Jayaprakash, a former Ph.D. scholar of Varanasi, is the CEO.

https://www.linkedin.com/in/vishnu-jayaprakash/

https://www.agzen.com/agzen-whatwedo

# AI solutions-Romesh Wadhwani, USA

Romesh Wadhwani is the Founder of the Wadhwani Foundation, a Silicon Valley entrepreneur, and a philanthropist. He built three successful companies, including the $3.5 billion Symphony Technology Group. A member of the Gates-Buffet Giving Pledge, he was appointed by President Barack Obama to the Board of Trustees of the John F. Kennedy Center. Romesh holds a B.Tech from IIT Bombay and an M.S. and Ph.D. in Electrical Engineering from Carnegie Mellon University. The Wadhwani Institute for Artificial Intelligence was established to develop and deploy AI solutions that benefit underserved populations in developing countries, a mission it has actively pursued since mid-2021. One such application is pest management for cotton farming.

## Technology

Bollworms cause an estimated 70% of all pest damage in cotton farming. The Indian government introduced Bt Cotton (Bollgard II seed), a genetically modified pest-resistant variety, to combat bollworm pests, including the American bollworm, pink bollworm, and spotted bollworm. However, the pink bollworm has developed resistance to Bt Cotton over time. Farmers traditionally monitor cotton fields with pheromone traps to detect pink bollworm populations, which guides their pest control measures. This manual data collection process, involving physical counting, analysis, and dissemination of information, is error-prone, unverifiable, and challenging to scale. In the USA, scientists have employed a different technique by breeding pink bollworms and using radiation to render them sterile. These sterile moths are then released over cotton fields to outnumber the wild moths, reducing their chances of finding fertile mates.

## Innovation

CottonAce is an AI-powered early warning system available as an app for Android smartphones. It helps farmers protect their crops by providing immediate, localized advice on the optimal timing for pesticide application. Lead farmers or extension workers use the app to upload photos of pests collected in pheromone traps. The AI model first checks the image for type and quality through an input validation module. It then identifies the pest type and isolates each pest by creating a bounding box around it. The model counts the pests within these boxes to determine if the number exceeds the Economic Threshold Limit (ETL), which indicates whether pesticide spraying is necessary. This information is shared with neighboring farmers, who do not require additional tools, such as smartphones. The app functions on simple smartphones and even works offline in areas with poor network coverage. It is available in nine languages: English, Hindi, Marathi, Gujarati, Telugu, Kannada, Tamil, Odia, and Punjabi. In addition to the app, a web-based dashboard supports program implementation by providing a real-time and historical view of pest occurrences in monitoring plots and villages.

## Paper

"Pest Management in Cotton Farms: An AI-System Case Study from the Global South": https://www.wadhwaniai.org/wp-content/uploads/2021/10/13_Pest-Management-in-Cotton-Farms-KDD.pdf

## Utilization

The primary users of the CottonAce app are lead farmers, chosen for their roles in guiding fellow farmers in their communities. Between 50,000 and 100,000 lead farmers grow cotton in India, trained through farmer welfare programs to manage local pest infestations, including with the aid of CottonAce.

https://www.wadhwaniai.org/rcats/pest-management/

# Bend-Insensitive Optical Fiber-Pushkar Tandon, USA

Pushkar Tandon graduated from IIT Delhi in Chemical Engineering in 1990. He earned his doctorate from Yale University in 1995 and completed his postdoctoral research at the University of Pennsylvania Institute of Medicine in 1998. Tandon is a 2022 inductee in the "National Inventors Hall of Fame" and is credited as the inventor on over 225 issued US patents and more than 350 patents worldwide. He has been a key inventor of several award-winning products, including those recognized by Time magazine as "Best Inventions of the Year," the R&D 100 Award, the IEEE Corporate Innovation Award, and the Indian Automotive Technology & Innovation Award.

https://www.linkedin.com/in/pushkar-tandon-1a7361b4/

## Technology

Corning had been exploring the use of optical fiber for residential applications, as running fiber directly into homes and offices could significantly enhance the speed and volume of data delivery. While early Fiber to the Home (FTTH) trials in Japan and France proved costly, widespread application awaited the development of passive optical networks (PONs). These networks use optical splitters to connect multiple subscribers on a single fiber, facilitating global deployment. Research intensified on bend-insensitive single-mode fiber to address limitations in traditional fiber installations.

## Innovation

Tandon joined Corning Inc. in 1998 and collaborated with colleagues to innovate optical fiber technology. Traditional optical fiber, which relies on light traveling in a straight line, was primarily effective for long-distance data transmission using straight cables. However, conventional fiber suffered significant signal loss when bent tightly, making it unsuitable for installations around indoor obstacles like walls or door frames. In 2004, Tandon and his team discovered that certain types of optical fiber could maintain efficiency around tight bends. They developed an optical "trench," a ring of material with a lower index of refraction, built into the fiber to reflect lost light back into its core. In regular graded index multimode fiber, multiple modes (or rays of light) travel down the fiber. While the inner modes are "strongly guided" and resistant to bending stresses, the outer modes are "weakly guided" and prone to loss when the fiber bends. Bend-insensitive fiber introduces a layer of glass around the core with a lower index of refraction, which reflects these weakly guided modes back into the core, preventing them from coupling into the cladding under stress.

## Patents

U.S. Patent Nos. 7,450,806; 7,903,917; 8,385,701

## Commercialization

Introduced by Corning in 2007, ClearCurve optical fiber can be bent to small diameters and navigate tight corners. This technology supports telemedicine, telecommuting, video streaming services, online gaming, and smart city technologies, meeting the growing demand for internet services, including cloud computing and data storage.

# Dropellette skin care-Madhavi Gavini, USA

Madhavi Gavini is the CEO and Founder of Droplette. She was also President and Co-founder of Novopyxis, a company with a patent-pending portfolio that includes novel therapeutics, drug delivery systems, and diagnostics, with a focus on engineering transdermal delivery devices for cosmetics and skin conditions. Gavini holds degrees from the Massachusetts Institute of Technology (MIT) and Johns Hopkins University.

## Technology

One of the most challenging aspects of skincare is ensuring the effective absorption of active ingredients through the skin. Retinol, a vitamin A derivative, is used in skincare to stimulate collagen and elastin production, which helps reduce fine lines, wrinkles, and hyperpigmentation. Collagen peptides, short chains of amino acids, can penetrate deeper skin layers and signal the skin to produce more collagen and elastin. Micropeptide technology, with an even lower molecular weight, further enhances collagen absorption. Glycolic acid is a chemical exfoliant that dissolves dead skin cells and oils, boosts collagen production, and maintains skin moisture without causing irritation from scrubbing.

## Innovation

Droplette works by creating a high-speed aerosol of tiny, rapidly moving droplets containing skincare actives that penetrate the skin barrier 20 times more deeply than topical treatments, without needles or pain. It is designed for use after cleansing and before moisturizing, in place of traditional serums. Droplette offers three skincare formulas: 0.15% retinol, 10.0% collagen, and 8.0% glycolic acid. Users select their desired active, insert a capsule into the device, and mist their face.

## Patents

"Fluid Delivery Devices and Methods," U.S. Patent No. 9700686, issued on June 3, 2016. The patent covers methods, systems, and devices for generating a mist, including a fluid delivery device with a housing, a fluid vaporizer that produces an aerosol mist, and a pump to accelerate the mist.

## Commercialization

Cofounded by Madhavi Gavini and Rathi Srinivas, Droplette was originally developed to deliver genes into the skin for treating rare skin diseases. However, the founders soon realized the technology's potential for beauty applications. The device uses a fast-moving aerosol mist to bypass the skin barrier, resulting in a more radiant, youthful complexion. The most popular Collagen capsules are infused with vitamin C, turmeric, and peptides to plump and brighten the skin. Glycolic capsules, featuring a blend of AHAs, BHAs, and PHAs, gently exfoliate and clear blemishes, while Retinol capsules combine retinol with collagen, arbutin, and rose oil to even skin tone and reduce aging signs without causing dryness.

For more information, visit: droplette.io/pages/our-story

# Driverless Tractor-Praveen Penmetsa, USA

Praveen Penmetsa is the CEO of Monarch Tractor and Founder/Chairman of Motivo. He graduated from Acharya Nagarjuna University and earned an MSME from the University of Cincinnati. Motivo develops intellectual property and builds transformative products for clients across energy, mobility, agri-tech, and aerospace sectors, serving a diverse range of clients from Global Fortune 500 companies to startups. The journey to the launch of Monarch Tractor began when Motivo Engineering, led by Praveen Penmetsa, won a USAID grant to develop and test the HARVEST tractor for emerging agricultural markets.

## Technology

In the 1950s, Ford developed a driverless tractor called "The Sniffer," but it was never produced due to its requirement of running wires underground through the fields. At the 2022 Consumer Electronics Show, John Deere announced the first autonomous tractor capable of operating without a driver in the cab. This technology is being integrated into their 8R 410 tractor, equipped with 12 stereo cameras and an Nvidia GPU, controllable via a smartphone.

## Innovation

Praveen Penmetsa collaborated with autonomous systems expert Zachary Omohundro and soon joined forces with manufacturing specialist Mark Schwager and viticulturist Carlo Mondavi. Chris Whitney, a seasoned technologist, later joined as head of autonomy. Designing a tractor presents unique challenges, more complex than designing an automobile, with a

12-speed gearbox and a hydraulic system for various farming implements, alongside preventing rollover on uneven terrain. Monarch Tractor, designed and built by this team, uses machine learning and AI to analyze data from onboard cameras and sensors, providing insights into pests, plant health, and other farming concerns.

## Patents

- Vehicle Rollover Mitigation, Publication Number: 20230249674

- Tight Turn Wheel Locking, Publication Number: 20230249687

- Three-Point Hitch Hook-Up Assist System, Publication Number: 20230247922

- Vehicle Lead Control, Publication Number: 20230249546

## Commercialization

Monarch Tractor's first Pilot Series tractor was deployed in April 2021 at Wente Vineyards, ahead of schedule. The company benefited from tools that accelerated its design and development and received support from the California Air Resources Board's FARMER program, aimed at reducing agricultural emissions. Taiwanese contract manufacturer Foxconn began production of the first MK-V all-electric tractors for Monarch at the Lordstown facility in Ohio. Following a technical partnership agreement, Indian tractor manufacturer VST Tillers Tractors Ltd invested $150 million for a 2% stake in Zimeno.

For more information, visit: monarchtractor.com/praveen-penmetsa

# Dynocardia-Mohan Thanikachalam, USA

Dr. Mohan Thanikachalam is a cardiovascular surgeon with over 15 years of experience and a leader in population-based cardiovascular disease (CVD) prevention. He is an affiliate faculty member at MIT, a research assistant professor at Tufts University School of Medicine, and a visiting professor at Sri Ramachandra University, Chennai. As CEO and co-founder of Dynocardia, he leads the development of ViTrack, a wearable, cuff-less device for continuous, non-invasive blood pressure (cNIBP) measurement. ViTrack provides real-time, beat-to-beat data on blood pressure, heart rate, rhythm, and respiratory parameters with the accuracy of an intra-arterial line, aiding in early intervention and chronic condition management. Dr. Thanikachalam holds an MBBS from Kasturba Medical College and completed his residency at Tufts University.

## Technology

Current blood pressure (BP) measurements typically rely on traditional cuff-based approaches. While ambulatory BP monitoring provides more continuous readings than office-based measurements, it is not suitable for frequent, repeated use. A true continuous BP measurement requires a cuff-less method that can be worn by the patient, with data stored electronically, similar to how heart rate and rhythm are monitored.

## Innovation

Dynocardia has developed a proprietary method to directly measure systolic and diastolic BP based on the spatiotemporal-force distribution on blood vessel walls and surrounding tissue. The company's optomechanical tactile sensor array can measure this distribution at the micron level. Using this technology, Dynocardia developed ViTrack, the first standalone, wearable, cuff-less, continuous, non-invasive blood pressure monitoring (cNIBP) device. ViTrack provides calibrated, continuous intravascular pressure waveforms, allowing for the measurement of central blood pressure, arterial stiffness, respiratory patterns, heart rate, heart function, arrhythmias, peripheral resistance, cardiac output, cardiac filling pressures, and other hemodynamic parameters.

## Patents

Tactile blood pressure imager, Patent number: 11883140

Opt mechanical Method to Measure Arterial Pulse and Assess Cardiopulmonary Hemodynamics, Publication number: 20230263414

## Commercialization

In 2015, with support from MIT and Tufts, the Dynocardia team began exploring a fundamentally different approach to blood pressure measurement that could operate automatically without user, caregiver, or healthcare provider participation. After extensive bench testing and ongoing clinical studies in the ICU, ViTrack technology has proven to provide accurate monitoring of systolic and diastolic pressure. The technology received the Breakthrough Innovation in Cardiovascular Digital Health HRX 2023 award from the Heart Rhythm Society and the American College of Cardiology Innovation Award in 2023.

For more information, visit: dynocardia.care/about

# Energy storage-Ramya Swaminathan, USA

Ramya Swaminathan is the CEO and founder of Malta, where she led the company's spin-off from X, Alphabet's Moonshot Factory (formerly Google X), in 2018 and guided it through multiple rounds of corporate venture funding and team recruitment. Before Malta, she was the CEO of Rye Development, a leading hydropower and pumped hydro storage developer in the United States. Swaminathan holds a Master in Public Policy from Harvard University's John F. Kennedy School of Government and a Bachelor in Anthropology from Amherst College. She advises the U.S. Department of Energy as a member of the Electricity Advisory Committee (EAC), serves as a DOE Clean Energy Education and Empowerment (C3E) Ambassador, and is a board member of Resources for the Future. The founding team of Malta includes Raj Apte, a prolific inventor with 85 patents.

## Technology

"Variable Energy Resources" (VERs), such as wind and solar power, are generation resources with outputs that cannot be perfectly controlled and are dependent on unpredictable fuel sources. As more solar and wind resources are integrated into the grid, power swings and instability increase. Long Duration Energy Storage (LDES) can address these issues by providing on-demand power and grid stability, optimizing renewable energy use. Thermal storage using molten salt is the most mature technology in this space.

## Innovation

Malta's thermo-electric energy storage system offers a flexible, low-cost, and scalable solution for storing energy over long durations at high efficiency. During charging, the system converts electricity into heat (stored in molten salt) and cold (stored in a chilled liquid), which can be stored for extended periods. When discharging, the system acts as a heat engine, combining the stored heat and cold to generate electricity. This dual-temperature approach makes the Malta system more efficient than other technologies.

## Patents

Adiabatic Salt Energy Storage, Patent Number: 11761336

Abstract: The system uses a working fluid in a closed cycle, involving a compressor and turbine, to efficiently exchange heat with hot and cold storage fluids. It operates as a heat engine or a refrigerator, using molten salt for hot-side storage and water for cold-side storage.

Inventor: Robert B. Laughlin

## Commercialization

Malta develops and operates utility-scale Pumped Heat Energy Storage (PHES) plants that, when combined with photovoltaic (PV) solar energy, provide reliable, emissions-free power overnight. Malta Iberia, the company's European affiliate, has secured a Project Development Assistance Agreement from the European Union and the European Investment Bank to develop a 100 MW, 10-hour duration energy storage facility in Spain.

For more information, visit: maltainc.com/company

# 75F IOT Based Smart Building Management-DEEPINDER SINGH, USA

Deepinder Singh, a graduate of Thapar Institute of Engineering & Technology, founded 75F in 2012 after designing some of the world's fastest core networks for Tier 1 service providers such as AT&T, NTT, and Verizon. With over 25 years of experience in electronics and computing, Singh developed 75F, an intelligent building solution leveraging the Internet of Things (IoT) and cloud computing to predict, monitor, and manage building needs.

## Technology

Traditional HVAC systems often rely on manual adjustments for peak heating and cooling loads, leading to occupant discomfort, inefficiencies, and labor-intensive maintenance. Smart building technologies, ranging from automated HVAC systems to security systems, aim to improve these conditions. 75F's solution builds on this legacy, using IoT and cloud-based technologies to optimize building environments.

## Innovation

75F's Dynamic Airflow Balancing™ system addresses HVAC temperature imbalances, enhancing comfort and energy efficiency. The Continuous Commissioning™ technology ensures HVAC systems operate only when needed, optimizing air quality and energy savings. Their end-to-end solution for central plant and chilled water systems integrates easily installed sensors and controls, providing multi-site and multi-system optimization across various heating and cooling equipment.

## Resources

https://www.75f.io/software/saffron-ai/

## Commercialization

75F offers a vertically integrated, full-stack solution that combines custom hardware and software to enhance comfort, air quality, and energy efficiency in buildings. Utilizing AI, 75F creates a digital twin of any commercial building, integrating data from a wireless sensor network and third-party weather forecasts to proactively manage air distribution. The IoT-native system stands out in the market for its connectivity, efficiency, and affordability.

In addition, 75F launched Saffron AI, the first AI platform (AIP) for commercial buildings. Saffron AI enables facilities managers to interact directly with building systems, leveraging real-time and historical data alongside industry standards to provide actionable insights. The platform uses 75F's thermostats and sensors to create a standardized digital twin, offering human-in-the-loop intelligence for energy management, schedule updates, and temperature adjustments.

For more information, visit: 75f.io/about-u

# IT Security-Jay Chaudhry, USA

Jay Chaudhry, born on August 26, 1958, is an Indian-American technology entrepreneur and the CEO and founder of Zscaler, a cloud security company. He graduated from IIT (BHU) and earned a postgraduate degree from the University of Cincinnati.

(https://www.linkedin.com/in/jaychaudhry/)

## Technology

Security Service Edge (SSE), as defined by Gartner, is a convergence of network security services delivered from a purpose-built cloud platform. It is a subset of the Secure Access Service Edge (SASE) framework, with a focus on security services. Core SSE services include Secure Web Gateway (SWG), Zero Trust Network Access (ZTNA), Cloud Access Security Broker (CASB), and Firewall as a Service (FWaaS). SSE supports remote workforces by providing consistent, enterprise-grade security and safe access, regardless of location. It eliminates data center backhauling, resulting in lower latency and improved performance, while allowing organizations to enforce uniform security policies, monitor user activities, and mitigate threats across their entire ecosystem.

## Innovation

As cyberattacks become more sophisticated and cloud applications more prevalent, traditional perimeter-based security models are increasingly inadequate. Zscaler's Zero Trust Exchange™ assumes that no user, workload, or device is inherently trustworthy. The platform verifies identity, determines destination, assesses risk through AI, and enforces policies before securely connecting users, workloads, or devices to applications across any network. Zscaler integrates NVIDIA AI technologies to enhance user experience and security-centric services. The use of NVIDIA AI capabilities, such as NIM inference microservices, NeMo Guardrails, and the Morpheus framework, improves data processing on the Zero Trust Exchange™ platform, enhancing its ability to defend against cyber threats and simplify IT and network operations.

## Patents

Zscaler's security offerings are protected by multiple patents in the U.S. and internationally, including Patent Nos. 7,899,849; 7,894,350; 7,921,089; and 7,984,102.

## Commercialization

Zscaler is a market leader with over 7,500 customers and analyzes more than 500 trillion data points daily across users, devices, networks, and applications. Leveraging AI and automation, Zscaler enhances visibility into sensitive data, provides contextual insights, and delivers closed-loop workflow automation. Zscaler's partnership with Google simplifies and strengthens Zero Trust security, integrating advanced threat protection with Chrome Enterprise and Google Workspace.

Zscaler Education offers a range of materials to support users in maximizing their familiarity with the Zscaler product suite.

**For more information, visit: zscaler.com/legal/patents**

# Integrated Innovation Systems-Suresh Sharma, USA

Suresh Sharma has advanced degrees and professional training in mechanical and aerospace engineering, business management, and global leadership from institutions including BITS Pilani, British Aerospace (UK), University of Florida (USA), and General Electric (GE). A former GE executive turned entrepreneur, Sharma exemplifies pragmatic business leadership and a deep understanding of commercializing innovations for lasting impact and economic growth.

Sharma has authored several best-selling business books, including:

➤ Energy 2040: Aligning Innovation, Economics, and Decarbonization (2024)

➤ Industrializing Innovation: The Next Revolution (2019)

➤ The 3rd American Dream: Global in Reach (2014)

➤ Global Outsourcing: Executing On-shore, Near-shore & Off-shore Strategy (2005)

His 2005 book on global outsourcing was a bestseller, sharing insights from his role at GE in the early 1990s, where he helped establish operations in over fifteen countries. The work is still used as a standard reference in the BPO industry and MBA programs worldwide.

## Professional services

Sharma serves as an Innovation Fellow at the Strategic Energy Institute and other interdisciplinary research institutes (IRIs) at the Georgia Institute of Technology (since 2020). He has also been an Industry Mentor at Flash Point, a startup incubator at Georgia Tech, and CREATE-X, a Georgia Tech program for entrepreneurs.

## Venture Funding

Having experience with top global industry and technology R&D organizations like GE Global Research, GE Energy, British Aerospace, NASA Langley, and Georgia Institute of Technology, Sharma founded Open Range Capital Partners in 2018. He has significantly contributed to several large smart city and iHUB projects, each valued at hundreds of millions of dollars, in locations such as Gainesville, FL; Oklahoma City, OK; Atlanta, GA; and Clarksville/Louisville, IN/KY. Recent ventures from this initiative include: Accure.ai (accure.ai), Cyberdome USA (cyberdomeusa. com), GridBlock (gridblock.com)

## Patents

During his time at GE Energy (1999-2000), Sharma was awarded a U.S. patent for an e-Business online application to optimize power plants (US Patent No. 7,275,025 B2). This innovation generated over $1 billion annually in online sales of upgrades for GE Energy. In total, Sharma and GE filed seven patents during his tenure.

## Cybersecurity

Cyberdome USA offers a one-day boot camp, "Intro to Cybersecurity for Managers," which provides practical cybersecurity skills using personal devices to help individuals and businesses stay safer online. For more information, visit: cyberdomeusa.com/cyber-academy.html

# LASIK Surgery-Rangaswamy Srinivasan

Rangaswamy Srinivasan spent 30 years at IBM's T.J. Watson Research Center, where he made significant contributions to laser technology. He holds 21 U.S. patents and has received numerous honors, including the National Medal of Technology and Innovation and the Russ Prize. Srinivasan earned his bachelor's and master's degrees in science from the University of Madras, India, in 1949 and 1950, and a doctorate in physical chemistry from the University of Southern California in 1956. He was inducted into the National Inventors Hall of Fame in 2002.

## Technology

The excimer laser, or exciplex laser, is a form of ultraviolet laser used in microelectronics, eye surgery, and micromachining. Since the 1960s, excimer lasers have been crucial in high-resolution photolithography for microchip manufacturing. In March 1973, definitive evidence of a xenon excimer laser at 173 nm was demonstrated by Mani Lal Bhaumik, a pioneer in laser technology.

## Innovation

Srinivasan developed a technique called "ablative photodecomposition," which uses laser pulses to remove tissue layers by breaking chemical bonds at a lower temperature, unlike previous methods that relied on heat and caused collateral damage. The excimer laser's ultraviolet light creates rapid temperature differences, vaporizing tissue and removing excess heat through convection, preventing burns to surrounding tissue. This precision made the excimer laser a potential replacement for traditional surgical tools. In 1983, ophthalmic surgeon Stephen Trokel collaborated with Srinivasan to apply this technique to corneal surgery, leading to the development of laser-assisted in-situ keratomileusis (LASIK).

## Patents

Excimer Laser Surgery: U.S. Patent No. 4,784,135

## Commercialization

Early LASIK procedures used a microkeratome blade to create a corneal flap and an excimer laser to reshape the cornea. In 1991, the automated microkeratome was invented, allowing for smoother cuts. In 1997, Juhasz and Ron Kurtz founded IntraLase Corp., which developed a bladeless LASIK technique using a compact instrument, further advancing the technology.

# Loka Chai Maker-Anish Sonpal, U.K

The Loka Chai Maker was promoted by Anish Sonpal, a legal professional with expertise in commercial contract negotiation, technology, outsourcing agreements, and telecom regulations. Co-founder Nikhil Sha is a creative and tech entrepreneur, an advisor, and an angel investor with a focus on consumer internet, creative industries, well-being, and consumer/D2C brands. Jon Marshall, a world-class industrial designer at the renowned design agency Pentagram, partnered with them to develop a solution for making chai at home without spilling.

## Technology

Traditional tea kettles are designed to boil water for brewing tea, with materials that can withstand high heat, such as metal, glass, or heavy ceramic for stovetop models. Teapots, in contrast, are used for steeping tea leaves in hot water and are made from more delicate materials like porcelain or ceramic. Unlike kettles, teapots are often decorative and not made to withstand direct heat.

## Innovation

The Loka Chai Maker is uniquely designed in the shape of an hourglass with an in-built filter to strain tea leaves and a curved rim to prevent spills. Its patented design creates a "chai fountain" effect when the mixture of water, milk, tea, and spices reaches a boiling point. Instead of boiling over, the chai circulates inward in an infinite loop resembling a fountain. This effect is achieved using a process called "baffles," which directs the flow inward rather than outward, creating the continuous "chai fountain."

## Patents

**https://www.pentagram.com/work/loka-chai-maker/story**

## Commercialization

The Loka Chai Maker project is currently seeking funding on Kickstarter. More details can be found at:

➤ kickstarter.com/projects/weareloka/loka-chai-maker/

➤ get.weareloka.com

For more information on chai startups from India and other funded projects, visit:

➤ businessalligators.com/best-leading-chai-startups-india/

➤ kickstarter.com/discover/advanced?sort=most_funded&woe_id=23424848

# Mash Makes-Biochar-Krishna Hara Chakravarty, Denmark

Krishna Hara Chakravarty is the Co-Founder and Head of R&D at Mash Makes A/S, focusing on Power2X, gasification, bio-coke, bio-coal, hydrogen, and biofuel projects. His expertise lies in agile management and carbon capture technology development.

## Technology

Biochar is a charcoal-like material produced from agricultural residues through pyrolysis, a thermochemical process that heats biomass without oxygen. Biochar is porous, allowing it to retain moisture and nutrients, improving soil structure by preventing compaction. This is particularly beneficial for clay soils, which lack air and water spaces. Adding biochar to soil helps keep it loose, aiding water retention and nutrient availability. According to the NRCS USDA, incorporating 10% biochar can reduce soil density to below 1.1 g/cm$^3$, ideal for root growth.

## Innovation

MASH Makes A/S was established in 2015 by Krishna Hara Chakravarty, Jakob Bejbro Andersen, Thomas Howard, Simon Strøm, and Jon Skovgaard-Petersen as a spin-off from the Technical University of Denmark (DTU). The company's pyrolysis technology produces carbon-negative biofuels and biochar from agricultural residues, such as cashew nut shells. The resulting bio-oil is refined into biofuels like biodiesel, reducing the fossil fuel footprint for sectors like shipping. Biochar production also sequesters carbon, making it a key technology for carbon dioxide removal and mitigating global warming.

## Patents

Mash Makes has one registered patent in the category of 'Petroleum, Gas or Coke Industries; Technical Gases Containing Carbon Monoxide; Fuels; Lubricants; Peat.' The technology platform, developed from knowledge at the Technical University of Denmark (DTU), is modular and containerized, allowing for scalable implementation. Several patents have been filed for this process.

## Commercialization

MASH Makes' second commercial pyrolysis facility is located near Udupi, Karnataka, India, adjacent to its first project financed by Nefco in 2022. The facility's prototype (4MW) and scale-up are manufactured by Enpro Industries, a leader in mechanical fluid systems and process equipment. One pyrolysis machine can service an area of 20-30 square kilometers in an agricultural belt. In October 2021, MASH Makes received a pre-purchase order for 6,000 tons of $CO_2$ Sink Certificates from the Swiss carbon removal marketplace, Carbonfuture.

For more information, visit: mashmakes.com

# Mati-earth, carbon removal-Shantanu Agarwal, USA

Shantanu Agarwal, the Founder and CEO of Mati-Earth, is a climate-tech entrepreneur with over 20 years of experience in the energy industry. He has successfully founded several companies in the climate-tech sector, including Susteon and Sustaera. His previous experience includes consulting with McKinsey & Co and working with Schlumberger and EV Private Equity. Agarwal holds an MBA from Harvard Business School and a B.Tech in Chemical Engineering from IIT Roorkee. Mati Carbon has been selected as one of the 20 finalists for the XPRIZE Carbon Removal grand prize of $50 million, to be awarded in 2025.

## Technology

Weathering is a natural process where rocks and minerals break down, reacting with CO2 in the presence of water to form bicarbonate ions, which are then washed into oceans for long-term carbon storage. Enhanced Rock Weathering (ERW) accelerates this process by applying crushed rocks, such as basalt, to agricultural land. Mati-Earth partners with smallholder farmers in the Global South to implement ERW as a scalable Carbon Dioxide Removal (CDR) strategy.

## Innovation

Basalt, a volcanic rock rich in silicate minerals, is ideal for ERW due to its abundance, cost-effectiveness, and ability to react with CO2. Its application in agriculture not only aids in carbon removal but also improves soil health by supplying minerals and correcting pH levels. As basalt dissolves, it releases cations like calcium and magnesium into groundwater, triggering a series of chemical reactions that convert CO2 in the soil to bicarbonate (HCO3–). This bicarbonate is then transported to streams, rivers, and eventually the oceans, where it stores carbon stably for over 10,000 years.

## White paper

For a detailed discussion on Enhanced Weathering (EW) and its policy implications, see the white paper: https://static1.squarespace. com/static/6054db4efc6c3622f12682fe/t/65faf8e2048ccf04bf556 1e7/1710946576798/EnhancedWeathering.pdf

## Commercialization

Mati-Earth uses advanced techniques, including Inductively-Coupled Plasma Mass Spectrometry (ICP-MS), to monitor basalt weathering in fields, detecting elemental concentration changes to calculate CO2 drawdown accurately. Mati Carbon is the first to scale Enhanced Rock Weathering (ERW) to an industrial level in India, operationalizing its model in Madhya Pradesh and Chhattisgarh, with plans to expand to three more states within the next year. The initial deployments in India also delivered the first engineered carbon dioxide removal credits.

For more information, visit: mati.earth/our-work

# Nano-Cybernetic Biotrek (NCB)-Dabina Sarkar, MIT

Prof. Deblina Sarkar is a trans-disciplinary innovator who began her training in electrical engineering and physics before expanding into biology. Her work is driven by the belief that disruptive technologies emerge at the intersection of diverse research fields. Her inventions include a 6-atom-thick channel quantum-mechanical transistor that overcomes power limitations, an ultra-sensitive label-free biosensor, a technology that reveals previously invisible biological nanostructures in the brain, and the first ultra-miniaturized antenna that operates inside a living cell in a 3D environment. Sarkar's PhD dissertation was recognized as one of the top three across the USA and Canada in Mathematics, Physical Sciences, and all Engineering departments.

Sarkar completed her undergraduate degree in electrical engineering at the Indian Institute of Technology (Indian School of Mines), Dhanbad. Her undergraduate research focused on nanoscale device design and spintronics, earning international recognition. Her 2007 paper explored double-gate MOSFETs, and she interned at Laurens Molenkamp's laboratory at the University of Würzburg, Germany, conducting research in spintronics. She graduated with her B.E. in 2008 and pursued both a master's and a Ph.D. at the University of California, Santa Barbara (UCSB).

## Technology

Nano-Cybernetic Biotrek (NCB) group at MIT has two major research directions:

Ddevelop novel nanoelectronic devices (such as Quantum Devices, Spintronics, Neuromorphic) employing ingenious device physics and smart nano-materials to achieve extreme energy efficiency, scalability and massive reduction of Green House Gases for sustaining the growth of Artificial Intelligence;

Merge such next generation technologies with living-matter creating unique nanomachine-bio hybrid systems, with remote control and wireless communication abilities to achieve unprecedented possibilities for probing/ sensing and modulating (for therapeutics) our brain and body.

## Innovation

Sarkar invented the Atomically Thin and Layered Semiconducting-Channel Tunnel-FET (ATLAS-TFET), a quantum-mechanical transistor that surpasses the power limitations of conventional transistors. This device achieves subthermionic subthreshold swing through quantum mechanical tunneling-based carrier transport, facilitated by a unique heterostructure design of doped germanium source, atomically thin $MoS_2$ channel, and a large tunneling area. Her other notable innovations include ultra-sensitive electrical biosensors, high-frequency models of graphene, and nanoscale brain mapping.

## Publications

https://web.mit.edu/deblina-sarkar/publications.html

## Technologies on offer

https://tlo.mit.edu/industry-entrepreneurs/available-technologies

# Orangewood labs-RoboGPT affordable robotics arms-Abhinav Kumar Das

Abhinav Kumar Das, a graduate of Guru Gobind Singh Indraprastha University, is the co-founder of Orangewood Labs (Y Combinator W18), along with Aditya Bhatia and Akash Bansal. Orangewood Labs focuses on developing affordable robotic arms for various applications. Aditya Bhatia studied industrial design, specializing in furniture at the National Institute of Design, Ahmedabad, while Akash Bansal is a finance professional. Previously, Das built trucks for rural areas in another venture.

## Technology

Orangewood Labs developed a cognitive programming framework that enables users to control robotic arms through voice or text commands in multiple languages. The system handles all edge cases, including safety, and adapts by learning about its environment. The framework uses custom-trained vision models and fine-tuned language models to facilitate interactions. Their robotic arm, nicknamed "Bro," operates with six degrees of freedom, similar to a human hand, and can lift up to 5 kg. The arm is available through a subscription model and can be programmed remotely using natural language commands.

For more information on building a voice-controlled robotic arm, visit: medium.com/@vardhanviswa/how-we-built-a-voice-controlled-robotic-arm-that-can-change-the-world-64af14d25d74 For further details on the development of a voice-controlled robotic arm, see: arxiv.org/abs/2303.09645

## Innovation

RoboGPT, developed by Orangewood Labs, is a live application for pick-and-place tasks using their robotic arm. The arm can be instructed using simple voice commands, such as "Bro, change the color from red to blue." This ease of use is made possible by a custom cognitive programming framework that continuously adapts to new environments and learns from interactions.

For more details, visit: ycombinator.com/launches/HeF-orangewood-democratizing-robots

## Patents

System and/or Method for Error Compensation in Mechanical Transmissions, US Patent No. 11148287B1, issued on October 19, 2021.

## Commercialization

Orangewood Labs, based in San Francisco and Noida, has partnered with Viam for data management and PickNik.ai for a software platform to develop robotics applications. Their business model offers robotic arms on subscription plans, making them affordable for small- and medium-sized businesses. Subscription costs range from $1,500 to $2,000 per month per robot, and the response in the U.S. market has been positive.

For more information, visit: orangewood.co/robogpt.html

# PECO Technology (Molecule air)- Dr. D Yogi Goswami

Dr. Dharendra Yogi Goswami, a graduate of Delhi College of Engineering, is a U.S. inventor, entrepreneur, author, and educator. He is a Distinguished Professor and Director of the Clean Energy Research Center at the University of South Florida. Goswami has published over 400 peer-reviewed articles, several books, and numerous book chapters. He is known for his contributions to renewable energy policy and the Goswami thermodynamic cycle. In 2016, he was inducted into the Florida Inventors Hall of Fame. Goswami co-founded Molekule, a company that markets an advanced air purifier.

## Technology

In 1995, Goswami proposed a thermodynamic cycle using an ammonia–water mixture as a working fluid, capable of producing both power and refrigeration from low-temperature heat sources. His innovations extend to air purification technology with Photo Electrochemical Oxidation (PECO), designed to reduce indoor allergens and pollutants. PECO removes particles as small as 0.1 nanometers by destroying organic matter not captured by traditional filters, including volatile organic compounds (VOCs), bacteria, viruses, mold, and other contaminants. PECO technology oxidizes organic matter on a nano-coated filter, breaking down pollutants into trace elements. This capability allows PECO to destroy particles 1,000 times smaller than those captured by HEPA filters.

## Innovation

Molekule's air purification technology, based on PECO, has been tested to inactivate up to 99.99% of viral particles, including H1N1 flu virus and coronavirus strains. Molekule's Air Mini and Air Mini+ have received FDA 510(k) clearance as Class II medical devices for home use to destroy bacteria and viruses, marking them as Molekule's first FDA-cleared products for medical use.

## Patents

Photocatalytic system for indoor air quality, Patent number: 5835840, Assignee: Universal Air Technology, Inventor: D. Yogi Goswami

## Commercialization

In 2014, Dr. Goswami, along with his son Dilip Goswami and daughter Jaya Rao, co-founded Molekule to commercialize the PECO technology. The Molekule Air Mini offers powerful air purification for smaller rooms (under 250 sq. ft.) using the same breakthrough PECO technology as the larger Molekule Air, effectively destroying pollutants in small spaces or throughout the home when used in tandem.

For more information, visit: molekule.com/products/air-purifier-air-pro

# Perplexity AI-Aravind Srinivas, USA

Aravind Srinivas, a dual-degree graduate from IIT Madras (IITM), is the co-founder and CEO of Perplexity AI, established in 2022 alongside Denis Yarats, Johnny Ho, and Andy Konwinski. The founding team brings diverse expertise: Yarats, the CTO, was an AI research scientist at Meta; Srinivas worked as an AI researcher at OpenAI; Ho, the Chief Strategy Officer, was an engineer at Quora and a quantitative trader on Wall Street; and Konwinski was part of the founding team at Databricks.

## Technology

Perplexity AI utilizes web crawlers, also known as spiders, to systematically browse and index the World Wide Web for relevant content. It combines this with a sophisticated Large Language Model (LLM) that processes vast amounts of text data to understand, interpret, and generate human language. This integration allows Perplexity AI to function as a conversational search engine, offering direct answers to user queries.

## Innovation

Perplexity AI distinguishes itself from other AI tools like ChatGPT by integrating multiple advanced models into its search functionality. It uses GPT-4o's vision capabilities for analyzing images, screenshots, and documents and employs Stable Diffusion XL, DALL-E 3, and Playground V2.5 for generating images from text prompts. For advanced tasks such as data analysis and detailed searches, Perplexity incorporates models like

GPT-4o, Claude 3 (Opus and Sonnet), and Sonar Large. Pro users have the option to select their preferred models within their account settings.

For more details, visit: explodingtopics.com/blog/perplexity-ai-vs-chatgpt

## Papers

For publications and academic work, visit: scholar.google.com/citations?user=GhrKC1gAAAAJ

## Commercialization

Launched in 2022, Perplexity AI is an AI chatbot-powered research and conversational search engine that provides answers using natural language predictive text, supported by web sources and citations. Operating on a freemium model, the free version utilizes Perplexity's standalone LLM with natural language processing (NLP) capabilities, while the paid version, Perplexity Pro, includes access to models like GPT-4, Claude 3.5, Mistral Large, Llama 3, and an Experimental Perplexity Model. As of early 2024, Perplexity AI boasts approximately 10 million monthly users.

Perplexity AI aims to become the world's most knowledge-centric company, backed by prominent investors such as Jeff Bezos, Elad Gil, Nat Friedman, Tobi Lutke, Jeff Dean, Susan Wojcicki, Yann LeCun, Naval Ravikant, Paul Buchheit, Andrej Karpathy, and others. It has developed the first generally available conversational answer engine that directly answers questions on any topic.

For more information, visit: perplexity.ai

# Point-of-Care Testing (POCT) Technology- Mukesh Kumar, U.K

Dr. Mukesh Kumar, a graduate with a BEng (First Class Honours) and a PhD in Physics from Lancaster University, is the co-founder and Technical Lead at Bailrigg Diagnostics Limited. He has co-authored numerous high-impact journal articles and filed two patents, with additional patents in preparation. Dr. Kumar was a finalist in the 2017 MIT Technology Review's 'Innovators Under 35 Europe' awards for his contributions to improving clinical blood testing technology. His research has been supported by grants from the Japan Society for the Promotion of Science (JSPS, London), Higher Education Innovation Fund (HEIF, UK), Engineering and Physical Sciences Research Council (EPSRC, UK), and Leverhulme Trust, among others.

For more information, visit: linkedin.com/in/dr-mukesh-kumar-1a696120/

Dr. Rashmi Mukesh, also a co-founder of Bailrigg Immunodiagnostics Ltd, holds a PhD from Lancaster University and is an alumna of BHU. She is developing a rapid healthcare assay for point-of-care testing (POCT) to provide quick measurements for conditions like acute myocardial infarction (MI), where rapid diagnosis is critical.

For more information, visit: linkedin.com/in/rashmi-m-780779135/

## Technology

Point-of-Care Testing (POCT), also known as near-patient or bedside testing, refers to medical diagnostics performed at or near the site of patient care. Lab-on-a-chip technologies, including glucose biosensor strips and lateral flow strips with immobilized antibodies, enable POCT by allowing bioassays like microbiological culture, PCR, and ELISA to be conducted at the point of care.

## Innovation

Bailrigg Diagnostics has developed a patented platform that integrates microfluidic processes, electroanalytical, and immunochemical sensors, with results interpreted by a handheld reader. This miniaturized technology provides quantitative results near the patient, allowing for immediate data access and interpretation. The results are wirelessly transmitted to custom apps and integrated into medical records, simplifying the diagnostic process and enabling rapid decision-making.

## Patents

Biomarker Sensor Apparatus and Method of Measuring Biomarker in Blood; Patent Number: GB1811279.7; Granted by the UK Intellectual Property Office in 2019.; PCT application filed in 2018 and currently undergoing national phases.

## Commercialization

Under Dr. Kumar's leadership, an entrepreneurial team called "eBIOGEN" was the first runner-up in an international business plan competition in India (2011) and was shortlisted for the finals at the 'RBPC 2012' business competition in Houston, known as the world's richest business competition. For more information, visit: bailriggd.com/Products-Pipeline.html

# Plant Based cheese-Debashree Roy, New Zealand

Debashree Roy is a Research Scientist and Agricultural Material Specialist at Riddet Institute, New Zealand. After graduating from Badan Chandra Krishi Vishwavidyalaya (BCKV) and earning an MSc at CFTRI, she joined Palmerston North (Manawatu), New Zealand, as a Research Scholar and now works at the Riddet Institute as a Research Scientist.

For more information, visit: linkedin.com/in/debashree-roy-9296a559/

## Technology

Cheese is a popular food globally, valued for its nutritional benefits, versatility, and taste. However, concerns over saturated fats, hormones in cow milk, animal welfare, and the environmental impact of dairy farming have driven demand for plant-based cheese alternatives. These alternatives often struggle to match the taste, texture, and nutritional value of dairy cheeses. Traditional plant-based cheese substitutes, which use vegetable proteins instead of casein, often result in impaired texture and functionality, including reduced elasticity, hardness, meltability, and stretchability.

To improve the quality of plant-based cheeses, manufacturers often include starches, gums, or gelling agents to mimic dairy protein functionalities. However, these additions generally result in a product with lower protein content (about 5-10%) than dairy cheeses (which have 15-30% protein). Consequently, plant-based cheeses frequently fail to meet consumer expectations for texture and flavor.

## Innovation

For the past four years, scientists at Massey's Riddet Institute have developed a fermentation process to create plant-based milk from legume seeds, leading to the creation of dairy-free creams and milk powders. The newly formed company Andfoods, backed by Massey Ventures and supported by Icehouse Ventures, has raised $2.7 million to commercialize these products. The researchers have developed a high-protein plant-based cheese product with properties similar to dairy-based cheese. The cheese contains about 5 wt% to 40 wt% protein and up to 10 wt% lipid, using pea and/or soy protein to achieve a total protein content of at least 40 wt%.

## Patents

Plant-Based Cheese Product, WO2023021428A1

Inventors: Debashree Roy, Harjinder Singh

This patent describes a plant-based cheese product with similar textural and sensory properties to dairy cheese and a process for preparing it.

## Commercialization

The plant-based cheese product is commercialized by Andfoods, co-founded by Arup Nag, Alejandra Acevedo-Fani, Yiran Wang, Debashree Roy, Harjinder Singh, and Alex Devereux.

For more information, visit: andfoods.co/about-us

# Sepsis Biomedical Sensor-Ambalika

Ambalika Tanak, MS'16, is a biomedical engineering doctoral candidate at the Erik Jonsson School of Engineering and Computer Science, University of Texas at Dallas. Her research focuses on developing biosensors to help physicians make faster decisions, potentially saving patients' lives. Her innovative work has led to scientific publications, a pending patent, and recognition as one of the top young medical researchers in the country. Dr. Shalini Prasad, head of the Bioengineering Department and the Cecil H. and Ida Green Professor in Systems Biology Science, is Tanak's advisor. Tanak's work on a sepsis detection device earned her a first-tier Baxter Young Investigator Award, which supports research in developing life-saving medical therapies and products.

## Technology

Sepsis is a severe response to infection that can result in tissue damage, organ failure, and death. It is particularly dangerous in COVID-19 patients, especially the elderly and those with pre-existing conditions. Currently, no rapid-testing method for sepsis exists. Tanak's sensor is a portable device that monitors five immune biomarkers using only a drop of blood plasma, providing a diagnosis within minutes.

## Innovation

Tanak and her co-authors developed a novel sepsis-testing sensor called DETecT Sepsis (Direct Electrochemical Technique Targeting Sepsis). This device offers the first proof of concept for rapid diagnostic screening of sepsis using a panel of immune response biomarkers. The DETecT Sepsis sensor showed high correlation with standard reference methods (Pearson's $r > 0.97$ for all five biomarkers) and achieved quicker response times with a simpler assay procedure. This was accomplished through a unique surface-engineered sensing strategy combined with an affinity biosensing principle.

For more details, visit: sciencedirect.com/science/article/pii/S0956566320307144

## Paper

https://www.sciencedirect.com/science/article/pii/S2590134619300258

## Commercialization

Drs. Shalini Prasad and Sriram Muthukumar are significantly involved in EnLiSense LLC, a company with a commercial interest in this research. Dr. Muthukumar, the CEO and Co-Founder of EnLiSense LLC, has secured over $1 million in non-dilutive funding from U.S. federal agencies for platform development and clinical validation. He has over 20 years of experience in electronic materials, device fabrication, and high-volume manufacturing, holding 17 granted patents and several pending applications. Dr. Shalini Prasad, a leader in the development of sweat wearables for disease tracking, was elected to the American Institute for Medical and Biological Engineering College of Fellows in 2022.

For more information, visit: enlisense.com

# SunDensity-Dr. Nishikant Sonwalker, USA

Dr. Nishikant (Nish) Sonwalkar is an inventor, entrepreneur, academician, and musician, known for his expertise in molecular dynamics of nanomaterials, photonic smart coatings, resilient solar microgrids, and brain-computer interfaces for adaptive learning. He has pioneered research in nano-optical coatings for photon conditioning in photovoltaic (PV) energy applications. Dr. Sonwalkar holds a U.S. patent for Photonic Smart Coatings with applications in photovoltaics, architectural glass, and optoelectronics.

Dr. Sonwalkar earned his doctoral degree from MIT in Molecular Dynamics of Nano-interfaces and Energy Materials and joined MIT as research faculty in Mechanical Engineering. He has also been recognized for his contributions to technology-enabled education, serving as the founding director of MIT's Hypermedia Lab and the Principal Educational Architect of MIT. His awards include the 2007 Innovative Excellence Award in Teaching and Learning and the 2007 USDLA Platinum Award for Best Practices in Online Distance Learning.

For more information, visit: linkedin.com/in/sonwalkar/

## Technology

In conventional solar panels, ultraviolet (UV) light typically converts to heat rather than electricity, leading to wasted energy and degraded solar cell performance over time.

## Innovation

Dr. Sonwalkar developed a Tandem Cell Technology using Perovskite Photovoltaics (PPV). Perovskites, known for their specialized molecular structure, efficiently absorb high-energy photons in the UV spectrum. In a solar module, the tandem active perovskite layer works with a silicon cell to convert more of the available solar energy across the wavelength range into electrons.

## Patents

Optical Coating for Spectral Conversion, Patent Number: 10935707

Abstract: The patent describes an optical coating with layers designed to reflect specific wavelengths and shift incident light to different wavelength ranges using nanoparticles. This coating enhances light absorption and conversion efficiency in photovoltaic applications.

## Commercialization

SunDensity is focused on developing advanced photonics solutions to enhance sunlight capture for clean, renewable solar power and optical instrumentation. Their Low-E Solar Control solutions reduce heat gain in buildings, lowering energy demand for environmental control.

For more information, visit: sundensity.us/low-e-technology

# Super conductors-Prof. Muralidhar Miryala, Japan

Prof. Muralidhar Miryala is a professor at Shibaura Institute of Technology, Japan, where he also serves on the Board of Councilors and is the former Deputy President. He is an OB Chair for the World Technology University Network and a professor at the Graduate School of Science and Engineering. Prof. Miryala graduated from Osmania University and has significantly contributed to the field of superconductors.

## Technology

In the early 20th century, low-frequency alternating current (AC) transmission was introduced in rail systems in Europe and the US, and this method is still in use today for most electrified main lines. Recently, there has been a shift toward direct current (DC) systems, which require shorter insulation distances and are particularly suitable for urban areas. DC power transmission for rail systems demands high current capacities, ranging from several thousand amperes in city centers to several hundred amperes in suburban areas.

## Innovation

Miryala's lab initiated a national project to develop a next-generation prototype DC superconducting cable for railway systems. The aim is to create a feeder for the overhead contact line system between substations and electric trains. Prof. Miryala invented a new class of mixed LRE-123 high-temperature superconducting system that operates up to 15 T at 77 K and at high temperatures up to 90.2 K. The team produced 30 and 300-meter Bi-2223 cables for trains operating at the Railway Technical Research Institute (RTRI). The project plans to develop DC cables ranging from 5 meters to 300 meters, with the goal of creating climate-friendly railways using DC superconducting cables to reduce $CO_2$ emissions.

## Patents

1. RE123 Based Oxide Superconductor and Method of Production

   a. US Patent No. 7964532 B2, Issued June 21, 2011

   Description: An RE123-based oxide superconductor with a conductive layer containing an $REBa_2Cu_3O_{7-\delta}$-based oxide superconductor, formed using a mixture of $RE_2BaO_4$ and $Ba_x$—$Cu_y$—$O_z$-based materials. The superconductor can use elements such as La, Nd, Sm, Eu, Gd, Dy, Ho, Er, Tm, Yb, Lu, and Y.

2. Oxide Superconductor of High Critical Current Density

   a. US Patent No. 6,063,736, Issued May 16, 2011

3. Seed Crystal for Production of Superconductor and Manufacturing Method

   a. JP Patent No. 2010-244160, Issued October 29, 2010

   For more information, visit: miryalalab.com

## Profiled in 2023

1. Acoustography- Dr. Jaswinder Singh Sandhu, USA

2. Bio surfactants-TeeGene Biotech. UK

3. Coal Science- Dr. Hardarshan Singh Valia, USA

4. Carrier Gas Extraction technology-Gradiant corporation, USA

5. Computer Science- Raj Reddy, USA

6. Flash Memory-Micron Technology, USA

7. Green Steel- Veena Sahazwalla, Australia

8. Nanocomposite Dental Materials- Sumita Mitra, USA

9. Regrowth of Bones- Nina Tandon, EpiBone, USA

10. Rotimatic- Zimplistic Pte, Singapore

11. Tiny Robots- Prof Vijay Kumar, USA

12. Water Technology Visionary- Anil Jha, USA

13. Wearable Sweat sensor- EnLiSense, USA

14. Wearable and reusable outpatient ambulatory ECG monitoring products- NimbleHeart, USA

# Agricultute (2024)

# AI driven operating system-Agam Khar/ Absolute, New Delhi

Agam Khar is the founder of Absolute, with Prateek Rawat joining later as a co-founder. Agam Khar has experience working at the Kalam Foundation and served as an advisor to Dr. APJ Abdul Kalam Technical University. Prateek Rawat, a graduate of the Institute of Engineering & Technology, previously worked as a Senior Consultant at Zinnov. Xenesis Institute, the R&D arm of Absolute, consists of a team with experience from Israel, the US, South Korea, and Africa. Dr. Prashant Khare, the Director of R&D at Xenesis, is a graduate of HS Gour Institute and has held a scientist position at the All India Institute of Medical Sciences in Bhopal. He completed postdoctoral studies at the Baylor Institute for Immunology Research and UT Southwestern Medical Center in Dallas, Texas, and has published over 30 research papers in journals like Nature Immunology and Nature Communications.

## Technology

One of the key technologies developed by Xenesis is BioAbling. Agricultural biologicals, which include microorganisms and biomolecules found in nature, are used to enhance plant growth, health, and productivity. Some biological products, such as biostimulants and biofertilizers, promote plant growth, while others, known as biocontrol agents, protect plants from bacterial or fungal infections and pests. Certain biologicals can simultaneously stimulate and protect plants, providing dual benefits.

## Innovation

Since its inception in 2015, Xenesis has developed several proprietary platforms. The Nature Intelligence Platform™ (NIP™) is a comprehensive database of microorganisms, secondary metabolites, signaling molecules, and other natural biomolecules. The AgFrontier group at Xenesis has developed the Signal Triggered Regenerative Activity Complex (STREAC), a technology that enhances the stability, shelf life, and effectiveness of farm biologicals. The biocare group introduced XenDHA, a novel solution for overcoming challenges in Omega-3 fatty acid supplements related to quality, yield, and contamination. Xenesis also developed Bio-Cat Insta Technology™ (BCIT™), which uses novel microbial strains to produce highly durable biocatalysts applicable in industries such as baking, brewing, detergents, fermented products, biofuels, pharmaceuticals, and textiles. Another tool, the Xenesis Observatory (XenO), utilizes Absolute's proprietary agri data stack to provide detailed insights into the health and productivity of smart farms.

## Patents/Papers

https://scholar.google.com/citations?user=OUmD9PwAAAAJ

## Commercialization

Absolute offers software solutions to control farm hardware systems, and collects data on a regularly basis from IoT devices, sensor suites, hardware systems, and satellite sources. It then feeds it into proprietary machine-learning algorithms to produce actionable insights. Absolute currently has a plant bioscience R&D platform BioX, a farm operating system (FARM OS), and a global trade platform for the produce. https://www.absolute.ag/

# Cryptoscope-IVRI, Izatnagar

Dr. Triveni Dutt is the Director of the Indian Veterinary Research Institute (IVRI) in Izatnagar. He graduated from Agra University and holds an MSc and PhD from ICAR. Since 2016, he has been a scientist at IVRI, focusing on animal breeding. Dr. Dutt is a Fellow at the Bioved Research Institute of Agriculture and Technology in Allahabad, India, and has an extensive research portfolio with 273 publications and 1,558 citations. researchgate.net/profile/Triveni-Dutt

## Technology

The goal of artificial insemination is to maintain a sufficient number of motile sperm in the oviductal isthmus at the time of ovulation to achieve the highest fertilization rate. The viable lifespan of sperm in the bovine female tract is difficult to quantify due to the heterogeneity of sperm cells, each with a different lifespan. Traditional heat detection aids include tail paint, Heatmount detectors, and the HeatWatch II System.

## Innovation

IVRI developed indigenous methodology, as a field tool for determining optimum time for fertile insemination in animals. The technology is based on crystallization pattern of cervical mucus. Bovine cervical mucus changes its biochemical composition and biophysical properties due to the variations in sex steroid levels during the oestrous cycle. As a consequence of oestrogen rise, cervical mucus is produced in larger amounts at oestrus—a stage also characterized by an increase in mucus crystallization when observed under light microscopy.

IVRI method-Place a drop of the cervical mucus from the animal in heat on the glass slide available with the Crystoscope. Prepare a thin smear. Allow it to dry for at least an hour or two. Now take this dried slide and place it in the designated slot of the instrument. Turn on the switch for the light inside the instrument, to see the fern pattern clearly. Three different patterns can be visible. If Nil pattern or an Atypical pattern is visible, do not take the animal for Artificial Insemination. Visibility of a Typical fern pattern is the sign of ovulatory heat and has maximum chances of conception if animal is inseminated during this period.

## Patents

All the patents of the institute are listed at:

https://www.ivri.nic.in/research/patents.aspx

Publication of the method: http://www.scielo.org.co/pdf/rmv/n28/n28a10.pdf

## Commercialization

The instrument is marketed under various names, such as HD Scope (Sidhant Trading), Lykascope (Lykya Exports), Labscope (Rohit Enterprise), Hariscope (Cattle Remedies India), and Ranscope (Ranbaxy India). Vamso, formerly known as Cattle Remedies Private Limited, was founded in 1969 by Dr. OP Agarwal (Veterinarian) and Dr. VP Agrawal (Ayurvedic Doctor).

# Driverless Tractor-Kaustabh Dhonde/ AutoNxt Automation, Thane

 Kaustubh Dhonde is the Founder and CEO of AutoNxt Automation Pvt. Ltd, a deep-tech start-up focused on developing advanced agricultural technology. The company is built on two core technologies: a high-torque electric powertrain and off-road automation. AutoNxt's primary product is an Electric Self-Driving Tractor, designed as the first application of these core technologies. Kaustubh Dhonde, a graduate of Dr. D. Y. Patil Vidyapeeth, successfully raised a seed round of Rs. 6.4 crore from investors including Chetan Maini, co-founder of SUN Mobility.

## Technology

Driverless tractors are categorized into two types: those with supervised autonomy and fully autonomous models. While companies like Mahindra and Monarch Tractors have introduced or planned driverless tractors in India, AutoNxt is focusing on developing a fully electric self-driving tractor. Their tractor employs an advanced electric powertrain and sophisticated automation software to perform agricultural tasks without direct human supervision.

## Innovation

AutoNxt's driverless tractor can be controlled via a proprietary software app. The farmer or owner inputs details such as the type of crop and task, and the tractor autonomously begins operations. Upon completion, the tractor notifies the farmer through the app. The app also provides real-time updates and alerts for any challenges, such as obstacles that the tractor cannot handle, like a large boulder or a fallen tree. The AI-based system allows the tractor to learn continuously, improving its navigation and task efficiency over time.

## Patents

AUTONOMOUS AGRICULTURE VEHICLE AND SYSTEM THEREOFAUTONOMOUS AGRICULTURE VEHICLE AND SYSTEM THEREOF, IN 201721028392 · Filed Aug 10, 2017.

## Commercialization

Driving a tractor in India can be as challenging as driving a car without shock absorbers on a rough road. Since building the first prototype in late 2017, AutoNxt has developed a fully electric drivetrain supported by a mobile app for live tracking and system monitoring. AutoNxt offers three variants of its autonomous electric tractor:

25 HP Variant: Equipped with a 15 KW motor and a 15 KWHr battery, it has a runtime of 5 hours on the farm (7 hours on the road) per charge and can carry loads up to 750 kg. This model is ideal for gardening, backyard, horticulture, and small-scale hauling. 35 HP Variant: Featuring a 25 KW motor and a 25 KWHr battery, it offers 6 hours of runtime on the farm (10 hours on the road) per charge and can carry loads up to 1,400 kg. Suitable for heavy farming tasks and medium-scale hauling. 45 HP Variant: Equipped with a 32 KW motor and a 35 KWHr battery, this variant runs for 8 hours on the farm (10 hours on the road) per charge and can carry loads up to 1,800 kg, making it ideal for large-scale farming and heavy-duty hauling.

# Fasal Kranti-Ananda Verma/Fasal, New Delhi

Fasal was founded by Shailendra Tiwari, a production engineering graduate from Varanasi, and Ananda Verma, a computer science engineer from Azamgarh. Shailendra, a graduate from NIFT, was part of the Startup Leadership Program's Bangalore cohort for 2019-20. Ananda Verma, with a background in engineering from VIT and an MTech from IIT, began his career at IBM Bangalore. Fasal aims to revolutionize agriculture with smart farming solutions.

## Technology

The core of modern smart farming involves sensors that provide data-driven insights and enable precision farming techniques. These sensors monitor, measure, and manage agricultural ecosystems in increasingly complex ways: Location Sensors: Use GPS to provide precise data on the spatial dynamics of fields and assets. Dielectric Soil Moisture Sensors: Measure soil moisture based on its dielectric permittivity, which is influenced by the water content. Optical Sensors: Assess crop health by using light waves to detect key physiological indicators. Biosensors: Equipped with biological components like enzymes or antibodies, they detect specific molecules or pathogens in plants or soil. Electrochemical Sensors: Monitor real-time environmental factors crucial for plant growth and soil health. The integration of data on temperature, humidity, and other factors with advanced analytics platforms allows farmers to forecast weather patterns, anticipate crop stress events, and make informed decisions regarding irrigation, pest management, and harvest timing..

## Innovation

Fasal Kranti is a solar-powered Internet of Things (IoT) system developed by Fasal. It is a plug-and-play device that can be assembled and deployed on a farm within 5 minutes, requiring no technical expertise. The system records various parameters at specific intervals using multiple sensors. The collected data is processed into actionable advisories using agronomic models, Artificial Intelligence (AI), and Machine Learning (ML). Farmers receive real-time alerts via the Fasal mobile app, providing farm-level, crop-specific, and crop stage-specific advice. Fasal Kranti is equipped with 12 different sensors that monitor various environmental and soil conditions, including canopy air pressure, air temperature, humidity, leaf wetness, light intensity, solar intensity, wind speed, wind direction, rainfall, soil moisture, and soil temperature, giving farmers a comprehensive view of their farm conditions.

## Patents

Patents are owned by Wolkus Technology Solutions Private Limited belonging to the same founders. Patent number 389766- System and Device facilitating determination of soil properties (2021)

## Commercialization

The system is installed in many farms, grapes and other horticulture products. https://fasal.co/case-studies.html

# Foldable Bulk Grain system-Mr. Ravindra Vithal Dekate, BMH Transmotion, Bhilai

Ravindra Vithal Dekate, a mechanical engineer who graduated from NIT Bhopal in 1985, has extensive experience in bulk material handling and mineral processing systems. Over 18 years, he worked at Bharat Heavy Electrical Limited (BHEL) in Bhopal and the Steel Authority of India Limited (SAIL) at the Bhilai Steel Plant (BSP). He also gained international experience in Nigeria, Jordan, Germany, and Canada (Worley Parsons, Vancouver). Leveraging this experience, Dekate founded BMH Transmotion in India in 2016, specializing in bulk material handling solutions.

## Technology

In India, a significant amount of grain is stored in open environments, leading to losses from pests, weather, and moisture, estimated at 8 to 12 percent. Traditionally, over 90 percent of grain storage has relied on 50-kilogram bags in warehouses or open-air Cover and Plinth (CAP) systems. Despite its inefficiencies, the open-air CAP system has been the standard for decades.

## Innovation

Mr. Dekate developed a patented, foldable, and portable hermetic (airtight) grain storage system called Foldable Bulk Grain Handling Technologies. This innovative system prevents pests from accessing the stored grain and eliminates the need for insecticides, enhancing grain quality. The system also monitors and regulates low temperature, moisture, and humidity, ideal conditions for grain storage that double the shelf life. A pilot project in 2017 in Dongargaon district, Gondia, Maharashtra, in collaboration with the Ministry of Tribal Affairs, demonstrated its effectiveness. Scientists from the Indian Council of Agricultural Research (ICAR) observed the system, noting a minimal loss of only 0.39%.

## Patents

➢ **Portable Grain Silo** Patent Number: 10462975

➢ **System and Method for Hermetic Storage of Agricultural Commodities During Shipping**

Patent Application Number: 20100270297

researchgate.net/publication/326697141_Foldable_Grain_Storage_Technology_A_New_Generation_Storage_System_for_Grain_Legume_Pulses

## Commercialization

BMH's Foldable Bulk Grain Handling Technologies offer a portable and foldable storage system that minimizes grain losses. The system can be deployed near farms where storage is limited, and losses are most significant. The storage bags are made from a specially developed fabric that is food-safe, reflects 60% of solar energy, is UV-resistant, temperature and cracking resistant, and fire and rodent retardant. bmhtransmotion.com/index.html

# Fortified Atta-Aishwarya Bhatnagar/ Greenday, Lucknow

Aishwarya Bhatnagar, Co-Founder of Greenday and Better Nutrition, is dedicated to empowering lives through nutritious choices and supporting holistic health and wellness. She holds a Bachelor of Science in Food Science & Nutrition from IHM, Mumbai. The company's Founder & CEO, Prateek Rastogi, earned his BA(Hons) from Sri Ram College of Commerce and completed his postgraduate studies at IIMA. The venture has attracted investments from World Champion PV Sindhu, who promotes better health through bio-fortified foods.

## Technology

Bio-fortification, first proposed in the early 1990s, is a sustainable strategy to enhance the mineral and vitamin content of staple food crops to address micronutrient malnutrition. This process improves the nutritional quality of food crops through conventional plant breeding, agronomic practices, and modern biotechnology. There are two primary approaches to bio-fortification:

Increasing Desirable Nutrients: Enhancing the content of nutrients such as provitamin-A, Vitamin C, protein, iron, zinc, calcium, lysine, tryptophan, anthocyanin, oleic acid, and linoleic acid in staple crops. Reducing Anti-Nutritional Factors: Lowering compounds like erucic acid, glucosinolates, and trypsin inhibitors. Indian agricultural institutes have developed 71 nutrition-rich crop cultivars, including wheat, rice, maize, pearl millet, finger millet, groundnut, mustard, lentil, small millet, linseed, soybean, cauliflower, potato, sweet potato, pomegranate, and greater yam. These bio-fortified varieties are also high-yielding, making them ideal for both nutritional and food security.

## Innovation

Greenday introduced several bio-fortified products, including Fortified Atta enriched with iron, zinc, and protein. Iron helps maintain healthy red blood cells and prevents anemia, zinc supports immune function and wound healing, and protein is crucial for fetal development during pregnancy and maintaining healthy blood cells. The company's innovation lies in establishing a supply chain for bio-fortified foods, promoting nutrition-dense farming, conducting soil testing for nutritional profiling, and using AI-generated soil health cards. These initiatives have encouraged widespread adoption of nutrition-dense farming and a gradual transition to organic farming through a phased replacement approach.

## Patents

There are patents covering the steps required to actually get a new gene into a plant. India does not grant patents for GM crops/ foods. In other countries also there are issues- as many as 16 patent and 72 intellectual property issues had to be resolved in the process of making Golden Rice available to poor farmers at no cost (http://www.goldenrice.org/).

## Commercialization

The products are available at Amazon and other e-commerce platforms. https://betternutrition.greenday.co/

# Forewarning of Plant Disease-Angshujyoti Das/Farmneed, Kolkata

Angshujyoti Das is the founder of Farmneed and the creator of Express Weather Service. His startup was funded by ICAR under the RKVYRAFTER scheme. Angshu is also a founding member of the Society of Satellite Professional International, India Chapter (SSPI).

## Technology

Agrometeorology has advanced significantly, enabling better short-term weather forecasting. Enhanced radar and satellite data, improved analytics, and advanced weather technology help farmers understand rainfall and growing conditions for specific fields. Agriculturally relevant variables such as active temperature sums, hourly precipitation, and cloud cover percentage are critical, but these metrics are not always accessible or accurate due to the limitations of models extrapolating data from meteorological stations. Data accuracy tends to decrease with lower model resolution.

## Innovation

Express Weather provides advanced weather information through forecasting technologies based on numerical models such as the Global Forecasting System and Weather Research and Forecast. The service customizes satellite data feeds based on geo-coordinates to provide tailored weather forecasts. The company's core components include:

➤ **Forecasting**: Provides short- and medium-range location-specific forecasts tailored to various needs.

- ➢ **Ground Observation**: Uses ground observatory units to monitor and record weather parameters.

- ➢ **Applications**: Develops user-friendly, weather-based applications that maximize economic gain and benefit different user segments.

- ➢ **Weather Stations**: Developed ATMOS, an automated weather station available in two versions—ATMOS1.1 with meteorological sensors and ATMOS1.2 with additional plant sensors. These stations are robust, sophisticated, and capable of transmitting data from remote locations via GSM or GPRS.

- ➢ **Dissemination**: Implements user-friendly processes to provide valuable information to beneficiaries.

## Patents

An example of related patent technology is the "Method for Building Up, Forecasting Methodology, and the Prediction Means of Weather Prediction Model" (CN107991722A/en) from China, illustrating the types of innovations being developed in this field.

## Commercialization

Farmneed is capable of making precise weather predictions within a 100-meter domain. Their 15-day predictions maintain around 90% accuracy for the first 10 days concerning rainfall, humidity, and dew point measurements. For corporate clients and organizations, Express Weather can install weather stations at a cost of Rs. 55,000 to Rs. 60,000 per unit.

For more information, visit: farmneed.com/about-us#intro

# Goodmylk-Plant-Based Dairy-Abhay Rangan, Bengaluru

Abhay Rangan founded Goodmylk at the age of 18 and is recognized as a Forbes 30 under 30 honoree for 2019 and 2020. He is also the founder of SARV (Society for Animal Rights and Veganism) and has been trained by Stanford University's D-School under the University Innovation Fellows program in Bangalore. In addition to his entrepreneurial efforts, Abhay is a TED Talk speaker, a podcast host on plant-based entrepreneurship, and a Carnatic vocalist. Goodmylk, which was later acquired by Nourish You, promotes plant-based dairy alternatives to make vegan food accessible to all. Notably, Abhay has been awarded PETA's 'Outstanding Activist of the Year' for his work in animal rights and veganism.

Nourish You was promoted by Rakesh Kilaru, Krrishna Reddy, and Sowmya Reddy and was the first brand to retail homegrown chia and quinoa in India. Investors include Jinisha Sharma, Aditya Agarwal, Victoria and Abhishek Shroff, and international backers like Sustainable Food Ventures (SFV) and VegInvest Trust.

## Technology

Plant-based milks have a rich history spanning different cultures over millennia. Coconut milk, almond milk, and soya milk have been popularized for their health benefits and adaptability in various recipes. The first soya dairy factory was established in 1910 by Li Yu-Ying, a Chinese biologist, who also patented the world's first soya milk. In the 1970s, under pressure from the dairy industry, plant-based milks were initially labeled as "liquid food of plant origin." Later, almond milk gained popularity, particularly in the USA, where its sales eclipsed soya milk by 2013. Rice milk, popularized by Rice Dream in the 1990s, was packaged in TetraPak cartons, increasing its availability and popularity.

## Innovation

The process of manufacturing plant-based milk substitutes involves breaking down plant materials to release oils and create simulated fat globules. For tofu production, soy milk is coagulated using various coagulants like calcium sulfate or magnesium chloride to separate proteins and fats. The choice of coagulant affects the texture and flavor of tofu. During coagulation, the soybean pulp thickens into curds, which can be controlled by adjusting coagulant concentration and stirring time.

## Patents

Dry fractionation for plant based protein extraction, US20160309744A1

## Commercialization

Goodmylk's products, sold under Nourish You's brand "One Good," include plant-based alternatives like mylk, curd, cheese, paneer, spreads, and butters. The plant-based food market has grown significantly in recent years, with vegan dairy being one of the fastest-growing sectors. This trend is driven by global pioneers like Oatly and the emergence of Indian startups such as MilkinOats (oat milk and chocolate), Sain (almond milk), and Strive by Plantbyte Foods (vegan protein beverages).

# Good Dot-Plant Based Meat-Abhishek Sinha, Udaipur

Abhishek Sinha is the Co-founder and CEO of GoodDot, a company specializing in plant-based meat alternatives. A chemical engineer from Pune University with a postgraduate degree in Law from NALSAR, Sinha transitioned to entrepreneurship after serving as Deputy Commissioner, IRS, in Udaipur.

## Technology

Plant-based meat has a history spanning over 2,000 years in China. However, the modern movement to establish it as a credible meat replacement began in the U.S. with companies like Beyond Meat and Impossible Foods leading the industry. In India, traditional meat substitutes like gluten (chakki), jackfruit (kathal), and more recently, soy nuggets and soy chaap have been popular among vegetarians and meat-eaters alike.

Plant-based meats are typically made from protein concentrates and isolates, combined with fats and starches. These ingredients are processed through methods like stretching, kneading, shear-cell processing, press forming, folding, layering, 3D printing, and extrusion to create meat-like textures. Two main methods for creating textured proteins are low moisture extrusion cooking (LMEC) and high moisture extrusion cooking (HMEC). Recently, a pilot-scale shear cell method for forming fibrous plant-based meat was developed by researchers at Wageningen University in the Netherlands.

## Innovation

GoodDot's UnMutton Keema is a plant-based alternative to traditional keema, requiring no refrigeration and boasting a 12-month shelf life without preservatives. The company's products are crafted from a blend of ingredients like soya flour, pea flour, quinoa flour, flaxseed powder, and rice flour, enriched with protein and essential micronutrients. These ingredients give the products a neutral taste that blends well with different spices and gravies, while providing a firm bite. Another popular product, Plant-Based BBQ Tikka, features protein chunks marinated in BBQ spices.

## Patents

While GoodDot does not hold specific patents, it operates in a field where major players like Impossible Foods hold significant patents. For example, Impossible Foods' patent (U.S. 9,700,067) was granted from a U.S. application based on a Patent Cooperation Treaty (PCT) application filed in 2014. The family stemming from that PCT application includes multiple granted patents and pending applications.

## Commercialization

GoodDot is a leading manufacturer of plant-based meats in India, offering a range of products including UnMutton Keema, Vegicken Curry Kit, Vegetarian Bytz, Proteiz, Pulao, Biryani, GoodDot Noodles, and My Secret Curry Paste. Plant-based meats generally have higher protein density (22-25g per 100g), zero cholesterol, and high dietary fiber compared to traditional meat, which contains no dietary fiber.

For more information, visit: gooddot.in/products/unmutton-keema

# Green Process for Enrichment and Isolation of UVA Absorber from Karanja Oil, Geemon Korah, Mane Kancor Ingredients Private Limited, Cochi

Dr. Geemon Korah is the Director and CEO of Mane Kancor Ingredients Private Limited, a leader in the field of global spice extraction and natural ingredients. Dr. Korah holds a graduate degree in agriculture, a postgraduate degree in business management, and a doctorate in business administration. He has also completed executive education programs at Harvard Business School, ISB Hyderabad, and IIM Calcutta. Since becoming CEO in 2006, Dr. Korah has successfully transformed Mane Kancor into a respected player in the natural ingredients segment and spearheaded a joint venture with V Mane Fils of France, one of the world's largest fragrance and flavor firms.

## Company Overview

Mane Kancor Ingredients Pvt Ltd., with over 50 years of experience, specializes in natural food ingredient solutions, encompassing sustainable sourcing, advanced research, extraction, formulation, testing, and delivery. The company operates in over 100 countries with multiple factories in India and regional distribution centers worldwide. Mane Kancor works closely with over 6,000 farmers who are Farm Sustainable Assessment (FSA) certified under the Sustainable Agriculture Initiative (SAI). The company is committed to environmental stewardship, supporting communities through corporate social responsibility (CSR) activities focused on health, education, and the environment. Mane Kancor's diverse product range includes oleoresins, essential oils, floral extracts, natural antioxidants, natural colors, culinary ingredients, ground spices, customized spice blends, and organic ingredients.

## Technology

Karanja oil (Pongamia pinnata) is known for its skincare benefits and use as a biofuel raw material. The key active component, Pongamol, is a photostable molecule that provides UVA absorption, making it advantageous in sunscreen formulations compared to other UVA absorbers. Traditional extraction methods have focused on solvent extraction to enrich or isolate Pongamol for cosmetic industry applications.

## Innovation

Pongamol content in Karanja oil naturally ranges from 0.3% to 0.6%, which was not viable for commercial isolation. Mane Kancor's researchers identified a homologue in Karanja oil and developed a green process to convert it effectively into Pongamol, increasing the content to 2.3% - 2.6%. This innovative conversion enhances Pongamol content by over 600%, making commercial isolation viable. Notably, the by-product oil after Pongamol isolation retains properties similar to the original Karanja oil, allowing it to be used for similar applications.

## Patents

Treated Karanja oil having increased Pongamol content and process for the preparation thereof, Indian Patent No.: 419944

## Commercialization

KanPro (Pongamol Min 95%) is one of the commercialized product variant from Mane Kancor Ingredients Private Limited.

# Imara-SML td, Mumbai

Imara (Technical: Fipronil 0.6% with Sulphur 70% and Zinc Oxide 13%) is a water dispersible granule (WDG) based product developed by SML Limited (formerly Sulphur Mills Limited), a global leader in innovative agricultural solutions. Established in 1971 by Mr. Deepak Shah, SML Limited specializes in crop protection, nutrients, soil health, and biologicals, serving over 80 countries worldwide.

## Technology

Water dispersible granules (WDGs) have long been used in agriculture, but they face challenges such as application difficulties and reliance on labor. When applied through mechanical means, such as hoppers and drillers, these granules often release at a single location and fail to distribute evenly. Additionally, biological materials like algae and bacteria, while useful as alternatives to chemical agents for improving soil nutrients, often fail to provide adequate nutrition and pest control throughout the crop lifecycle due to leaching losses. The challenge remains to develop an agrochemical or crop nutrient that can be easily applied, provides immediate uptake, and sustains its effects throughout the entire crop cycle.

## Innovation

Imara addresses the issue of yellow stem borer, a pest that can reduce paddy yield by up to 80% in late-planted crops. It is a unique insecticide and nutrient combination that replaces inert ingredients with Sulphur (70%) and Zinc Oxide (13%), which are active components beneficial for both pest control and plant nutrition. This innovative formulation not only protects plants from harmful inerts but also enhances soil health, living up to its tagline: "Imara Ek, Faayde Anek" (One Imara, Many Benefits).

Research on Imara's effectiveness was conducted by SML's R&D team and validated by institutions such as Visva Bharati University, Annamalai University, and Konkan Agricultural University..

## Patents

### Agricultural Compositions

Patent Number: US20180325105A1

Inventors: Thankapan Vadakekuttu, Arun Vitthal Sawant

Current Assignee: Sulphur Mills Ltd

This patent covers a water disintegrable granular composition that includes at least one agrochemical and one or more agrochemically acceptable excipients. The composition is designed for effective uptake by plants and soil, providing consistent release and protection throughout the crop cycle.

## Commercialization

Imara is designed for ease of application in paddy crops. It is recommended to apply 4 kg per acre, 15-20 days after transplanting in paddy or 25-30 days after sowing in direct-sown crops. The product acts as a protective shield against stem borer infestations, ensuring better yield and crop health.

For more information, visit: sml-ltd.com

# Intello Labs-Fruit Sorter-Milan Sharma, Gurugram

 Milan Sharma is the CEO and Co-Founder of Intello Labs, a company that focuses on AI-driven solutions for agriculture and food processing. An IIT Bombay graduate, Milan gained experience in big data analytics and machine learning at companies like Snapdeal, Evalueserve, and AbsolutData. Co-Founder Himani Shah, also from IIT Bombay with a dual degree in mechanical engineering, now leads strategy at Intello Labs after co-founding Allied Nanotechnologies Universal. Nishant Mishra, another IIT Bombay alumnus, is responsible for technological innovations, architecture, and development. Co-founder Devendra Chandani, a graduate from Nagpur University based in New York, drives global sales. The startup was incubated at the JioGenNext accelerator, where it had the chance to conduct a pilot with Reliance Retail.

## Technology

Machine vision technology has become a transformative force in agriculture and food processing, especially in the classification and grading of fruits. Effective sorting and categorizing of fruits are essential for maintaining quality, enhancing supply chain efficiency, and minimizing food waste. This technology employs various methodologies. Color analysis uses color cameras and algorithms to extract color features from images, aiding in fruit classification. Contour detection and geometric analysis are applied to extract critical data about the size and shape of fruits. Meanwhile, texture analysis evaluates the surfaces of fruits for smoothness and roughness, contributing to quality assessment.

## Innovation

Intello Labs has developed the AI-driven defect detection platform, Intello Track, which brings several innovative features to the agricultural sector. The platform uses advanced AI algorithms to accurately identify and classify defects in fruits, ensuring that only the highest-quality produce reaches consumers. It provides real-time processing, optimizing the sorting process and reducing delays in production. The system is highly adaptable, capable of working with a variety of fruits, making it a versatile solution for diverse agricultural needs. It also enhances quality assurance by addressing imperfections and inconsistencies to meet the highest industry standards. Moreover, Intello Track offers data-driven insights into sorting analytics, enabling informed decision-making and process optimization for increased productivity.

## Patents

A system and method for grading agricultural commodity, WO2020208540A1, Inventor- Nishant Mishra

## Commercialization

Intello Track can assess up to 40 types of commodities. The Intello FruitSort system uses a unique transport mechanism to handle fruits gently, preventing damage while sorting by weight and detecting external defects with high accuracy. This technology surpasses traditional methods by identifying imperfections not visible to the naked eye, ensuring superior quality control.

https://www.intellolabs.com/

# Lab Grown Meat-Siddharth Manvati/Clear Meat/Delhi

Dr. Siddharth Manvati is the Co-Founder and CEO of Clear Meat, India's first lab-grown meat company. A biotechnology graduate from Dr. DY Patil College, he completed his Ph.D. at Shri Mata Vaishno Devi University and conducted extensive research in synthetic biology at Jawaharlal Nehru University (JNU). Alongside his ventures Clear Meat Pvt Ltd and Foresight Biotech Pvt Ltd, Dr. Manvati has focused on developing safe, nutritious, and affordable slaughter-free meat alternatives. Clear Meat aims to revolutionize the food industry by offering lab-grown meat as a sustainable option. The company recently developed ClearX9, an alternative to Fetal Bovine Serum (FBS), as a general-purpose medium for growing animal cells based on naturally available resources. Foresight Biotech, another spin-off from the research at JNU, focuses on designing novel biotherapeutics, enzymes, and other biomolecules for applications in medicine and biotechnology.

## Technology

The concept of lab-grown meat gained prominence in 2013 when Dr. Mark Post of Maastricht University unveiled the world's first cultured beef burger. This innovation has since inspired a global movement towards cultured meat, with startups like Aleph Farms (Israel), Finless Foods (USA), Higher Steaks (UK), and IntegriCulture (Japan) emerging as key players. Lab-grown meat involves cultivating animal cells in a controlled environment to produce meat without the need for traditional livestock farming.

## Innovation

Clear Meat's groundbreaking innovation, ClearX9™, is an alternative to Fetal Bovine Serum (FBS), commonly used in cell culture. FBS is a nutrient-rich fluid essential for cell growth, but its production involves the slaughter of millions of pregnant cows each year, raising ethical and environmental concerns. ClearX9™ offers an affordable, sustainable, and ethical substitute for FBS, using natural substances instead of animal-based serum. It has demonstrated effective results in growing various cell lines, including HeLa (cervical cancer cells), HEK293T (embryonic kidney cells), and Nthy Ori-3-1 (thyroid epithelial cells), maintaining good cellular health and growth patterns. ClearX9™ shows promise as a credible alternative to FBS, with potential applications in toxicology testing, biomanufacturing, regenerative medicine, and vaccine research.

## Patents

Clear Meat holds a patent for a supplement composition used in cell culture (Publication number: 20240034986). The patent describes a natural supplement composition for growing animal cells without the use of animal-based serum, providing a slaughter-free, effective, and affordable solution for cell culture. The inventors listed are Pawan K. Dhar and Siddharth Manvati.

## Commercialization

ClearX9™ is positioned not only for use in the cultured meat sector but also for broader applications in biotechnology and pharmaceutical industries, as well as in university labs and research institutes. Currently, ClearX9™ is undergoing product testing in several academic institutions, marking a significant step towards commercialization and wider adoption in scientific research and development. For more information, visit clearmtl.com.

# Lab Grown Sea Food-CMFRI/Neat Meatt Biotech, Kochi

The Central Marine Fisheries Research Institute (CMFRI) has entered into a collaborative research agreement with Neat Meatt Biotech, a startup focused on lab-grown seafood. This public-private partnership, signed by CMFRI Director A. Gopalakrishnan and Neat Meatt Biotech co-founder and CEO Sandeep Sharma, aims to advance research in the field of cultured seafood. Sandeep Sharma, a seasoned bioprocess engineer with nearly 20 years of experience in R&D and manufacturing, holds a Ph.D. from Aligarh Muslim University (AMU) and has worked extensively on vaccines in various biotech companies. His expertise in both upstream and downstream processing is instrumental in developing novel technologies for cultured meat production.

## Technology

Lab-grown seafood, also known as in vitro or cultured seafood, is produced by cultivating cells taken from aquatic animals under controlled lab conditions in a bioreactor. This process creates a mass of muscle and fat cells without involving bones, scales, or other non-edible parts. The technique mirrors developments by companies like BlueNalu, Wildtype, and Avant Meats, who are pioneering cell-based seafood such as fish fillets and crustaceans. The process starts with collecting cell samples from fish or shellfish and then growing them into muscle tissues in bioreactors. This method eliminates the need for traditional fishing or aquaculture, presenting a sustainable and cruelty-free alternative to conventional seafood.

## Innovation

Producing cell-based meat (CBM) on an industrial scale requires unique cell lines tailored for agricultural applications. Establishing a stable cell line is a significant challenge in CBM, as these lines must be used repeatedly for meat production cycles. Some companies, such as Kerafast (USA) and ESCO ASTER (Singapore), already manufacture commercial cell lines of various species. In India, the ICAR-National Bureau of Fish Genetic Resources maintains a repository of cell lines from 50 fish species, including the Bluegill fry and Rainbow trout. For the development of lab-grown seafood, researchers are manipulating stem cell markers in culture media to induce myogenesis, the process of muscle formation. Additionally, databases like the Hilsa Transcript SSRdb provide valuable genetic data to support the development of unique cell lines for lab-grown seafood

## Patents

The publication "Cell-based Meat: An Emerging Paradigm in Fish Meat Production" discusses the technology and processes involved in lab-grown fish meat production. This publication can be accessed at krishi.icar.gov.in

https://krishi.icar.gov.in/jspui/bitstream/123456789/82871/1/Cell-based%20Meat%20An%20Emerging%20Paradigm%20in%20Fish%20Meat.pdf

## Commercialization

Neat Meatt Biotech is currently in the development phase of its cell-based seafood products. The firm aims to introduce a range of cultured seafood options to the market, providing a sustainable alternative to traditional seafood. The development of cultured seafood aligns with the company's broader goals of promoting cruelty-free and environmentally friendly food sources. neatmeatt.in/about-us

# Methane Bio-Conversion-Ezhil Subbian/ stringbio.com, Bengaluru

Ezhil Subbian is the Chief Executive Officer and co-founder of String Bio Pvt Ltd, a company dedicated to methane bio-conversion technologies. Ezhil holds a degree in Industrial Biotechnology from Anna University and a Ph.D. in molecular biology and biochemistry from Oregon Health and Sciences University. She founded String Bio initially in the USA, later relocating it to Bengaluru, India. Under her leadership, the company has received numerous accolades, including winning $100,000 at the Future Food Asia Awards in 2017 and innovation grants from various government authorities. String Bio has also achieved recognition in global competitions such as the Hello Tomorrow Deep Tech Global Startup Challenge in Paris and the Liveability Challenge in Singapore. Vinod Kumar serves as the Co-Founder and Managing Director.

## Technology

The concept of methane-eating microbes for fermentation dates back to the 1980s when Statoil in Norway first explored the technology. This innovation was commercialized by Aptagen in 2000 and later by Calysta in 2011. String Bio builds on these advancements with its proprietary String Integrated Methane Platform (SIMP). This platform leverages methane, extracted from natural gas or biogas, as a feedstock to produce high-value products. Through a unique biotech-hardware approach, SIMP incubates methane-fed bacteria to produce alternative protein powders, which can be used across various industries, including animal feed, cosmetics, textiles, and agriculture.

## Innovation

String Bio's innovation lies in its ability to convert methane into valuable products through a sustainable and carbon-negative process. The company's patented solution, the String Integrated Methane Platform (SIMP), employs a proprietary fermentation technology that converts methane into protein-rich ingredients. These ingredients have a robust amino acid and mineral composition optimized for animal nutrition, providing health benefits such as improved gut health and better weight gain. String Bio's first protein powder product has been tested extensively in animal farms and agricultural research centers across India, Asia, Australia, the US, and the EU, demonstrating its efficacy and market potential.

## Patents

String Bio holds over 55 patents globally related to biology, fermentation, and engineering. One notable patent is for their processes for fermentation and purification of value-added products from gaseous substrates (Patent number: 10883121).

## Commercialization

String Bio targets the poultry and aquaculture sectors for its methane-derived protein powder, which has already received approval for use in animal feed in the EU. The company is backed by investors, including Srinivasa Hatcheries. In addition to animal feed, String Bio produces bio-stimulants for agriculture, such as CleanRise, which enhances crop tolerance to abiotic stress, improves nutrient utilization, and boosts soil fertility and microbiota. CleanRise has shown to reduce methane emissions by up to 50% and nitrous oxide emissions by up to 40%, while increasing crop yield by up to 39%. www.stringbio.com/Agriculture.html.

# Milk Analyzing platform-Rajat Pandya/ FaunaTech, Bengaluru

Rajat Pandya, CEO and Founder of FaunaTech, is a serial entrepreneur with a history of starting successful ventures such as Square Foods, Pepperchic, and Custominch. A graduate of the University of Mumbai, he also attended the Stanford Seed Program. FaunaTech was part of the cohort 4 of the AgFunder GROW Impact Accelerator and is incubated at C-CAMP in Bengaluru, India. C-CAMP, in collaboration with Beyond Next Ventures (BNV) from Japan, is working to create the C-CAMP – BNV Innovation Hub (CBIH), focusing on life sciences and biotechnology startups. This initiative aims to facilitate Japanese investment in Indian startups, particularly in deep science innovations. FaunaTech has been recognized with the International Dairy Federation - Dairy Innovation Award 2023 for 'Innovation in Sustainable Farming Practices - Animal Care'.

## Technology

FaunaTech focuses on innovative solutions to improve dairy farming practices, particularly in the early detection of mastitis, a condition that affects about 25% of milch animals worldwide. Traditional methods like the California Mastitis Test (CMT) provide a low-cost, on-farm diagnostic tool for detecting subclinical mastitis. However, newer technologies like mass spectroscopy can provide more precise measurements by identifying changes in milk constituents, such as somatic cell counts (SCC) and mastitis-causing pathogens.

## Innovation

FaunaTech has developed "Fauna," a smartphone-based, handheld diagnostics platform designed for early detection of critical diseases, monitoring cattle health, and screening milk quality directly at the farm level. The platform aims to reduce antibiotic usage by allowing farmers to detect mastitis in cows and buffaloes within five minutes, offering the precision of blood diagnostics to the dairy industry. The device measures over 10 key parameters from raw milk, including somatic cell count, fat content, antibiotic residue, and BHB levels. It uses proprietary optics and spectral techniques to provide real-time, accurate results. The technology is currently ready for pilot deployment and serves as an end-to-end solution that assists cooperative managers and farmers in testing, analyzing, and managing care for their livestock.

## Patents

FaunaTech has filed a patent titled "A PORTABLE DEVICE FOR TESTING AGRO-DAIRY BASED SAMPLES" (Publication number: 20240192200). The patent describes a portable device equipped with an optical detector based on micro-spectroscopy. The device can accommodate various testing methods, including a colorimetric test strip, a concave-shaped holder for solid or semi-solid substances, and a cuvette.

## Commercialization

FaunaTech's product, based on proprietary sensing technology that supports dry chemistries, lateral flow tests, and direct spectral measurements of various fluids, is currently undergoing testing. The device's ability to provide rapid, on-site diagnostics is set to revolutionize the dairy industry by enabling better disease management and improving milk quality control.

www.faunatechsolutions.com.

# NECTAR POUCH-Harini/Nutrigenetics life Science, Nilgiris

Er. Harini R is the Chief Operating Officer at Nutrigenetics Life Science Pvt Ltd, a budding entrepreneur recognized for her innovative contributions and recipient of the prestigious BIRAC SIIP Fellowship. With a passion for palm trees and their challenges, Harini has been instrumental in developing solutions that enhance the collection and preservation of Neera, also known as palm nectar. She completed her engineering graduation from Avinashilingam University and pursued an MTech at the National Institute of Food Technology, Entrepreneurship and Management (NIFTEM-T). Harini has published research on the bioactive components of traditional South Indian fruits and their potential health benefits in chronic diseases.

## Technology

Neera, or palm nectar, is a sap extracted from the inflorescence of various palm species. It is a sweet, translucent liquid highly susceptible to natural fermentation at ambient temperatures, often fermenting into toddy within a few hours post-extraction. Neera is widely consumed in countries like India, Sri Lanka, Africa, Malaysia, Indonesia, Thailand, and Myanmar. It does not require mechanical crushing or leaching, unlike sugarcane or beetroot; instead, it is obtained by slicing the spathes of palms and scraping the tender part below the crown. Several technologies have been developed to process and preserve Neera in its natural form, retaining its vitamins, sugars, and other beneficial nutrients. The National Chemical Laboratory in Pune and the Central Food Technological Research Institute in Mysore have developed specialized filtration and preservation techniques to extend Neera's shelf life.

## Innovation

Harini has developed the "Necollect" pouch, a specialized collection bag designed to enhance the efficiency and hygiene of Neera collection. The Necollect pouch incorporates chill-trapping technology to prevent fermentation, maintaining a stable temperature during collection and ensuring the sap remains unfermented and fresh. This innovative pouch is reusable, washable, and break-resistant, providing a sustainable alternative to conventional plastic bags used in palm sap collection. By reducing the carbon footprint and utilizing renewable resources for chilling, Necollect offers a cost-effective and environmentally friendly solution. It also features properties that resist insect intrusion and protect against climatic variations, making it an ideal tool for collecting palm sap under hygienic conditions.

## Patents

Harini has filed a patent for her innovation titled "Nectar pouch for collecting unfermented nectar from palmyra palm," with Application No: IN 202341066055.

## Commercialization

Ready to market.

# Pesticides E-Sticker-Barnali Ghatak/ Xolligent, Kolkata

Dr. Barnali Ghatak is the Founder and CEO of SBH Electrocloud and Xolligent. She is a highly accomplished researcher with a Master of Technology in Nanotechnology from NIT Durgapur, where she was a Gold Medalist, and a Ph.D. from Jadavpur University, Kolkata. She currently leads SBH Electrocloud Pvt. Ltd. and Xolligent LLP while serving as an INSPIRE Faculty Fellow at Aliah University, Kolkata. Dr. Ghatak's research focuses on developing advanced sensing technologies, such as Molecularly Imprinted Polymer (MIP) based Quartz Crystal Microbalance (QCM) sensors for fruit maturity assessment, among other innovative projects.

## Technology

Conventional methods for detecting pesticide residues are often time-consuming and impractical for on-site or field use. Among the various approaches, including electrochemical and optical sensing, colorimetric sensing is emerging as a promising solution for the on-site detection of pesticide residues. This method offers quick detection, easy fabrication, high selectivity, and sensitivity, and is readable by the naked eye, making it highly suitable for rapid field analysis. Recent advancements in this field include the development of solution-based nanoparticles, portable label-free inverse opal photonic hydrogel particles, and paper-based analytical devices like chip-based colorimetric assays and smartphone-coupled assays.

## Innovation

Olligent has developed a colorimetric sensor for pesticide detection that utilizes the enzymatic interaction between acetylcholinesterase (AChE) and indoxyl acetate (IDA). The presence of pesticide molecules, such as Chlorpyrifos, inhibits the enzyme's activity, preventing the formation of a blue pigment. In practical terms, the circular paper in the colorimetric sensor remains white if the immobilized enzyme is completely inhibited by the pesticide, indicating contamination. Conversely, the paper turns blue if the enzyme is not completely inhibited, signifying a lack of pesticide presence. This innovative approach allows for a simple, yes/no swab-based detection method that is non-destructive and highly effective for assessing pesticide residues on agricultural produce.

## Patents

A Method For Non Destructive Colorimetric Detection Of Pesticides,' Inventors: Soumyo Mukherji, Tathagata Pal, Patent Application No.: 202021024457

## Commercialization

SBH Electrocloud, in collaboration with Xolligent, is developing a range of innovative products, including conductive paint, Food e-Tester, customized Screen Printed Electrodes, capacitive strips, and molecularly imprinted polymer sensors. These products are designed to evaluate qualitative biomarkers at the time of agricultural produce collection or harvesting, ensuring digital traceability throughout the value chain from farm to fork or farm to pharm. The Pesticides E-Sticker and other sensing systems developed by the company aim to enhance food safety and quality assurance in the agricultural and dairy sectors.

https://sbhelectrocloud.com/about/

# Portable Instant Quality Assessment Solution-Padmini Sampath/GREEN COLLAR AGRITECH SOLUTIONS, Bengaluru

Padmini Sampath, the Co-founder and CEO of Green Collar Agritech Solutions, holds a degree in Electronics and Communication Engineering from the Government College of Technology, Coimbatore. With over a decade of senior leadership experience in electronics and semiconductor development, Padmini has partnered with Sarangan Thirumavalavan, the Co-founder and CTO of Green Collar Agritech Solutions. Sarangan, a graduate of NIT-Trichy in Electrical and Electronics Engineering, brings over 20 years of experience in semiconductor product development and system design.

## Technology

Assessing the quality of agricultural produce is a significant challenge, particularly in India, where standards such as the Indian Standard document (IS 3576:2010) specify a minimum curcumin content of 2.0% on a dry basis for turmeric. Traditional methods like wet chemistry and chromatographic techniques, although accurate, are expensive, time-consuming, and prone to human error and manipulation. Non-destructive methods such as Fourier-transform infrared (FTIR) and near-infrared (NIR) spectroscopy are essential for analyzing low concentrations because they preserve the sample's integrity, providing high sensitivity and the capability to detect chemical compounds at trace levels. For instance, the 'Raman 1.0', a hand-held device for detecting adulteration in edible oils, showcases the practical application of such technologies. This device, developed by Oak Analytics, can test over 250 samples per battery charge and stores data on the cloud, marking a significant advancement in food testing.

## Innovation

Founded in 2022, Green Collar Agritech Solutions focuses on AI and ML-based solutions for quality assessment, particularly in the agricultural, nutraceutical, and pharmaceutical value chains. The startup developed TARAM®, a portable, cloud-connected quality assessment tool designed to provide reliable, repeatable results. The TARAM® device, weighing less than 280 grams, features an intuitive user interface that enables anyone to use it effectively. It offers rapid results in under two minutes, is cloud-connected for immediate data access, and is available at a fraction of the cost of traditional testing methods.

## Patents

An ergonomic enclosure for handheld portable spectrometer, 202441037709

An additive manufactured gasket for sealing between interfacing physical geometry, 202441033885

A stochastic noise reduction system through iterative model feedback, 202441055051

A system for dynamic management of pre-processing, execution of machine learning models on mobile app, 202441055052

## Commercialization

The TARAM® product has been commercially launched for assessing the quality of turmeric powder, blends, and rhizome. The company is currently conducting pilot programs with major extractors, processors, and Farmer Producer Organizations to expand its market reach and application scope. greencollar.ai/index.html.

# Plant Based Cheese-Anushi Patel/Soft Spot Foods, Mumbai

Anushi Patel, the founder of Soft Spot Foods, is an industrial designer from the University of Arts, Philadelphia. Her background in industrial materials like plastic, metal, and wood, combined with her passion for sustainable design, has significantly influenced her approach to food innovation. She has previously worked on projects like a collapsible pizza box, a thesis on food waste in America, and bioplastics made from cornstarch. Patel was mentored by renowned chefs Alex Sanchez and Divesh Aswani, and her venture, Soft Spot Foods, won the "Best Vegan Cheese" award by PETA India in 2020.

## Technology

The production of plant-based cheese analogs has seen a significant rise in recent years. According to the Good Food Institute, plant-based cheese sales grew by approximately 42% from 2019 to 2020, reaching $270 million in the USA alone in 2020. This surge is notable given that a decade ago, plant-based cheeses were often considered inferior to their dairy counterparts in terms of taste and nutritional value. The challenge in creating plant-based cheese lies in replicating the complex physicochemical, sensorial, and nutritional properties of traditional cheese. Regulations in many countries also restrict the use of dairy-related terms for plant-based products, adding another layer of complexity to the industry. Innovations in this field include the work of Dr. Debashree Roy from Massey University, who developed new technology for sustainable plant-based foods, leading to the creation of the startup ANDFOODS.

## Innovation

The advancements in vegan cheese alternatives over the past decade have largely been driven by ingredients such as cashew nuts, known for their creaminess. Cashews are blended into a paste, fermented, and then mixed with various ingredients to achieve different flavors, such as truffle or chili. Similar to dairy cheese production, some vegan cheesemakers allow fermented cashews to age, depending on the product. For example, cream cheese is not aged, while other types are left to mature for up to two weeks. Soft Spot Foods has developed a range of plant-based cheeses, including Soft Mozzarella, Cheddar, Parmezan, and Truffle Cheese Spread, using innovative ingredients like cashew nuts and watermelon seeds to create textures and flavors that mimic traditional cheeses.

## Patents

Soft Spot Foods holds a patent for their unique plant-based cheese composition and the method of making it: "Plant-based cheese composition and method of making," Publication number: 20230309574.

## Commercialization

Soft Spot Foods has successfully brought its range of plant-based cheeses to the market, catering to the growing demand for dairy-free alternatives..

https://softspotfoods.com/product/truffle-spread/

# Satellite Images for Agriculture-Prateep Basu/SatSure, Bengaluru

SatSure is a decision analytics company headquartered in Bangalore, India, with additional offices in Philadelphia, U.S., and Liverpool, U.K. The company was co-founded by Prateep Basu, Abhishek Raju, and Rashmit Singh Sukhmani. Prateep Basu, the CEO, is an aerospace engineer from the Indian Institute of Space Science and Technology (IIST) and has a certification in Smart Space from Oxford University. With experience at ISRO, he leads SatSure and its subsidiary KaleidEO. Abhishek Raju, the Chief Growth Officer, is a graduate of Bangalore University with a certification in understanding rural India and impact insurance. Rashmit Singh Sukhmani, the Chief Technology Officer, is an IIST graduate with an MBA from Thompson River University, who also has a background with ISRO and Wikistrat. SatSure leverages satellite technology, machine learning, and big data analytics to provide insights across agriculture, infrastructure, and climate action. The company also collaborates with academic institutions, including IIT Roorkee.

## Technology

Satellites have been used in agriculture for decades, with programs like Sentinel, the Copernicus program, and Landsat leading the charge on the commercial front. However, the sector is currently experiencing disruption from new players offering data processing and analytical solutions. One of the primary challenges in utilizing satellite data is its format—it is often not easily consumable, requiring experts to process and interpret the data.

SatSure is addressing this gap by developing systems that can transform raw satellite data into actionable insights for agriculture and other sectors.

## Innovation

SatSure has developed a patented system and method for processing satellite imagery with high-frequency revisits to monitor vegetation using a deep learning model. The system involves one or more satellites that capture a series of images over time. The vegetation monitoring server receives these satellite images and adjusts them using pre-processing techniques to align different spectral bands. A deep learning model then reconstructs images by replacing cloudy and shadowed pixels with the corresponding pixels from another set of images, effectively generating clear, usable data for vegetation monitoring.

## Patents

System for producing satellite imagery with high-frequency revisits using deep learning to monitor vegetation, Patent number: 12039452 Inventors: Prateep Basu, Rashmit Singh Sukhmani, Sanjutha Indrajit

## Commercialization

Services offered:

CROP INSURANCE-Reduce your losses with smarter risk estimates and real-time monitoring.; AGRICULTURE TRADING-Find accurate production and acreage estimates for any plot of land.; AGRICULTURE BANKING-Improve loan management with relevant data and analysis.

AGRICULTURE INPUTS-Understand customer needs with smarter sales intelligence.

FOOD PROCESSING-Get total visibility into every stage of your supply chain.

https://www.satsure.co/agriculture.html

# Silage baling machine-Madhav Kahtira/ Cornext.in, Hyderabad

Madhav Kshatriya is the CEO & Managing Director of Cornext Agriproducts Private Limited. He holds an MBA from Australia with expertise in sales, strategy formulation & management. He has over 5 years experience only in Agritech & Dairy apart from his decade experience in other sectors. For our pioneering effort in bringing Baled Silage to Indian dairy industry, we have been awarded National Best Startup Award 2020 by Govt. of India.

## Technology

Silage, a fermented animal feed, was introduced in the late 1800s, and can also be stored in a silage or haylage bale, which is a high-moisture bale wrapped in plastic film. These are baled much wetter than hay bales, and are usually smaller than hay bales because the greater moisture content makes them heavier and harder to handle. These bales begin to ferment almost immediately, and the metal bale spear stabbed into the core becomes very warm to the touch from the fermentation process. Silage or haylage bales may be wrapped by placing them on a rotating bale spear mounted on the rear of a tractor. As the bale spins, a layer of plastic cling film is applied to the exterior of the bale. This roll of plastic is mounted in a sliding shuttle on a steel arm and can move parallel to the bale axis, so the operator does not need to hold up the heavy roll of plastic. The plastic layer extends over the ends of the bale to form a ring of plastic approximately 12 inches (30 cm) wide on the ends, with hay exposed in the center.

## Innovation

The company developed mini Silage baler with features like: Patented Baling Technology, Lowest cost of baling – 5 paisa /kg Silage, Produces 50kg bales which is most suitable for easy handling & usage, Compaction Rate – 650 kg/m³; ideal rate for best quality silage, 6 layered wrapping with UV protection. Daily Output – 20 to 25 MT, Automatic bale wrapping, weighment & ejection.

Video: https://youtu.be/f6McpQ14G0M?si=U5cV9daE2RBI6QIa

## Patents

**Patent no- 201741045240, date of filing 15/12/2017,**

## Commercialization

The machine is sold in India and also exported to Brazil, Tanzania, Nepal. Thailand etc.

https://www.cornext.in/mini-silage-baler-msb-500/

# Silk Spinning Cum Reeling Machine (Mulberry Silk, Muga Silk and Eri)-Nabajit Bharali, Guwahati

Nabajit Bharali isa prolific innovator from Assam. He received National Awards from the then Hon'ble President of India for his contribution to grassroots innovations in 2017.

## Technology

Unnati, Solar Silk Reeling + Spinning Machine (Patented) is World's first compact and portable Solar Powered Silk Yarn Reeling + Twisting machine. Ideal for reeling of Tassar & Muga Silks. Yarn for both Warp and Weft can be produced on Unnati.

https://www.reshamsutra.com/machine/unnati

## Innovation

This Silk Spinning Cum Reeling Machine is an important device (machine) for those engaged in sericulture and those working as weavers. As a significant number of rural women in Assam are engaged in the sericulture and weaving sector, therefore, this machine with special and unique features will expedite their work in less expensive manner.

Advantages of the machine developed by Nabajit Bharali.

1. This reeling and spinning machine will help in the reeling and spinning of various types of silks (eri, muga, mulberry silk etc.) in less time as compared to reeling and spinning machines available in the market. 2. The spinning machines available in market can spin 800 cocoons in eight (8) hours; however, this Automatic Reeling and Spinning Machine that he has innovated can spin 6000 to 7000 cocoons in eight (8) hours. 3. An amount of thread that requires one hour to be reeled manually, the same amount can be reeled in 2 to 3 minutes only through this Automatic Reeling and Spinning Machine. 4. This machine runs on (operates by) electricity or battery and it is rechargeable. 5. The bobbin of this Automatic Reeling and Spinning Machine will stop automatically as soon as the sinning/reeling work gets completed. 6. This Automatic Reeling and Spinning Machine is less expensive (affordable for rural women). The Spinning Machine available in the market is priced higher (around 55,000); however, this Automatic Reeling and Spinning Machine is less expensive (around 1500).

Video- https://youtu.be/DOu4rjavcaA?si=YB411gnLIBEThOvm

## Patents

Patent No: 487662.

## Commercialization

Contact innovator for 3 innovations:

(a) Automatic Body Controlled Wheelchair.(Patent No:158211)

(b) Silk Spinning Cum Reeling Machine (Mulberry Silk, Muga Silk and Eri). Patent No: 487662.

(c) Automatic feeding Machine for upper limb immobilised challenged people.

Nabajit Bharali.S/O:Rudra Bharali. P.O &; Vill: Michamari,P.S: Gogamukh., State: Assam, India. Dist: Dhemaji.Pin: 787026, Email Id:nabajitdmj@gmail.com,Phonenumber:+917002588251,+919613723935

# Solar Powered Insect Traps for Farmers- Durgaprasad/AEDAA Equipment/ Madanapalle

Durga Prasad is the CEO of AEDAA Equipments Pvt Ltd, a company he co-founded in 2022 with Murali Thota. The startup, incubated at Siddaganga Incubation Foundation in Tumakuru and supported by AIC PECF at Pondicherry Technological University and N.G. Ranga University (RKVY Rafter), aims to revolutionize farming with innovative technology solutions. Durga Prasad, a graduate from Sir Vishveshwaraiah Institute of Science & Technology in Madanapalle, is focused on developing advanced tools to assist farmers in managing their operations more effectively. AEDAA Equipment's flagship product, the Farm Assistant Model, integrates irrigation automation with disease and pest prediction capabilities, providing a comprehensive solution for modern agriculture.

## Technology

Traditional sticky traps, commonly used in agriculture, are colored paper or plastic boards coated with glue that attract and capture insects. Different insect species are attracted to different colors, and these traps are designed to exploit this behavior. In addition to sticky traps, UV fly light traps are another method used to capture flies. These traps attract flies using UV light and capture them on glueboards when they come near. The effectiveness of these traps can be significantly enhanced by integrating them with a light source.

## Innovation

AEDAA Equipment has developed a **Solar-Powered Pest Controller**, which is ideal for a variety of crops and storage facilities such as go-downs. This device leverages solar energy, making it an environmentally friendly solution that does not rely on external power sources. The pest controller features advanced UV attraction technology, using slightly visible UV light to attract flies and other insects. These pests are then trapped in a tank filled with soapy water and oil located beneath the light. The device operates continuously on solar power, supplemented by a built-in lithium-ion battery with a Battery Management System (BMS), which ensures energy efficiency and reduces operational costs. This approach minimizes the need for harmful chemicals and pesticides, promoting healthier crop growth and a safer environment.

## Patents

Pest trap with UV light, though a simple product, has several patents. They address issues like power from multiple LEDs, positioning of LEDs, avoiding radiation outside target area etc. Example is this Korean patent:

https://patents.justia.com/patent/10827738

## Commercialization

Various products offered by the firm are:

PVC Remote Gate Valve for Precision Water Control. Available in 2-inch, 2.5-inch, and 3-inch sizes, our PVC valves cater to a variety of irrigation needs, ensuring a perfect fit for your existing farm setup. Operating on solar power complemented by a Li-ion battery with a built-in Battery Management System (BMS), these valves offer an eco-friendly and cost-effective solution, reducing dependence on external power sources.

https://www.aedaa.in/

## Profiled in 2022

1. Nano Urea-Dr. Ramesh Raliya and IFFCO

2. AI-IoT Robotic Aeration Device for aqua-ponds- Bariflo Labs.

3. Poultry Litter Raking: Pakshi Mitra Poultry Technologies

4. Automatic Composting Machine - Agnys Waste Management.

5. Biofertilsers- BomLife

6. Farm Mechanization- IndusTill FarmTech

7. Smart Electric Tiller- SANDBIRD

8. Precision Drip irrigation- MAGBIITY

9. Crop Specific, Microbial Bio-products - LCB Fertilizers.

10. Soil Testing- Neoperk Technologies

11. Cattle Health Monitoring- Flixdrop Technology

## Profiled in 2023

1. 4G NANO" Based Nutritional Agri In-Puts- ICAR & Pratishta Industries, Hyderabad

2. Agri PV-RenCube, Bangalore

3. Anti-Pesticide topical gel- Praveen Kumar Vemula lab at inStem, Sepio Health Private Limited, Bangalore

4. Bio-Stimulants- BioPrime AgriSolutions Pvt Ltd, Pune

5. Digital Moisture Analyzer- A-GRAIN, Ambala

6. DME Tractor- TAFE, IIT Kanpur

7. Food processing plants- Best Engineering Technologies, Hyderabad

8. Fruit bar from prickly pear fruits (opuntia ficus indica)- Dr. Chenna Kesava Reddy Sangati, IIPM, Bangalore

9. GroTron autonomous irrigation system- Farmagain Agro Pvt Ltd, Coimbatore

10. Innovative Rhizome Processing-S4 foods, Aurangabad, Maharashtra

11. Kappahycus alvarezii elite seedling production- Dr. M. Ganesan, CSIR-CSMCRI, Gujarat

12. Machine to remove insects from stored grains- Professor (Retd) Mohan Sriramasarma, Tamil Nadu Agriculture University (TNAU), Coimbatore,

13. Multipurpose Processing Machine- Dharamveer Singh Kamboj, Yamuna Nagar, Haryana

14. Non-invasive and non-intrusive honey harvesting device- KLE Technological university, Hubballi

15. Optical fruit sorting system- Zentron Labs pvt ltd, Bangalore

16. Onion warehouse with IOT- Godaam Innovations, Nashik

17. Production of seedlings in agarose yielding red seaweed Gracilaria dura- Dr. Vaibhav A. Mantri, CSIR-CSMCRI

18. Pesticidal water dispersible granule formulation- Parijat Industries (India) Pvt. Ltd, New Delhi

19. Smart soil monitoring system- Proximal Soilsens Technologies Pvt. Ltd, Pune

20. Supply chain management for FPOs Vesatogo Innovations, Nasik

21. Solar Insect Traps- SAFS Ecotech Pvt Ltd, Puducherry

# Healthcare (2024)

# A Multi-Position Wheel Chair-Neomotion/IITM

NeoMotion Assistive Solutions, co-founded by Swostik Sourav Dash, a BTech and MTech graduate from IIT Madras, is a startup incubated at the Indian Institute of Technology Madras (IITM). The startup is closely associated with the TTK Center for Rehabilitation Research and Device Development (R2D2) at IITM, headed by Prof. Sujatha Srinivasan, who is a Professor in Mechanical Engineering. R2D2 focuses on research and development of assistive and rehabilitation devices for individuals with movement impairments, supported by initiatives like IMPRINT-1 (MHRD and ICMR), HDFC CSR, and Tata CSR.

## Technology

Many wheelchair users, particularly in rural areas, face significant challenges with their existing mobility devices. Traditional wheelchairs often cannot fit inside their homes and are unsuitable for outdoor use on rough terrain. Currently, there is no single mobility solution that caters to both indoor and outdoor needs. Typically, separate products—a wheelchair for indoor use and a tricycle for outdoor use—are required, making it difficult and unsafe for users to transfer between devices.

NeoMotion's solution to this problem is **NeoRider**, an innovative attachment that transforms a standard wheelchair into a tricycle within seconds, enabling users to navigate rough terrain safely and effortlessly. This attachment was initially designed to be compatible with popular wheelchair models, but further research and development revealed that existing wheelchairs were not robust enough for prolonged outdoor use with the NeoRider. This insight led to the development of a more durable and versatile wheelchair, **NeoFly**, and its motorized outdoor counterpart, **NeoBolt**..

## Innovation

NeoStand-India's most customizable and compact Electric Standing Wheelchair - NeoStand. This innovative mobility solution, created in collaboration with NeoMotion, is specifically designed for individuals with limited arm strength, featuring a motorized standing function activated at the push of a button. The wheelchair's compact design facilitates maneuverability in constrained spaces, while its extensive customization options ensure a perfect fit for the unique requirements of each user. NeoFly is the first Indian Wheelchair that is customised to the user. It is built to add higher comfort, higher mobility, higher manoeuvrability and higher ergonomics. NeoBolt is the ultimate innovation that helps Wheelchair Users move around independently, not within the house, but to Work, to Shop, to Anywhere. All the user needs to do is, to Strap on a front scooter. The user can do this independently without anyone's help. Move around the city independently.

## Patents

EASY TO USE PORTABLE MANUAL STANDING WHEELCHAIR WITH SAFETY FEATURES AND FOR OUTDOOR USE, Publication number: 20200078237, Inventors: Sujata Srinivasan, Vivek Sarda, Sourav Swostik Dash

## Commercialization

https://www.neomotion.in/

# AI-Radiology-Amit Kharat/Deeptech, Pune

Dr. Amit Kharat, a clinical radiologist with extensive experience in musculoskeletal imaging, is the co-founder of DeepTek.ai, an AI-driven healthcare radiology diagnostics startup. Dr. Kharat holds degrees in medicine (MBBS, DMRD, DNB) and has a Ph.D. in Musculoskeletal Imaging. He has also pursued further education in hospital administration and has a certification from MIT Management School on AI's implications for business strategy. Dr. Kharat co-founded DeepTek.ai in 2018, leveraging his expertise in radiology and emerging technologies. His research includes multiple indexed publications in the field of medical imaging. The startup is also co-founded by Ajit Patil, a serial tech entrepreneur with a background in IT services and strategic equity alliances, and Ashutosh Pathank, CTO, who has over 20 years of experience in the technology industry.

## Technology

The widespread use of electronic health records (EHRs), medical imaging archives, and annotated datasets has provided ample training data for AI algorithms. At the same time, improvements in hardware, such as graphical processing units (GPUs) and distributed computing, have accelerated the deployment of computationally intensive AI models.

## Innovation

DeepTek.ai has developed several AI-driven tools to enhance radiological diagnostics. One of their flagship products is the **DeepTek Chest Pathology AI Analyzer**, designed to assist healthcare practitioners in identifying chest pathologies from chest radiographs. This tool helps improve the accuracy and speed of chest pathology detection, facilitating worklist prioritization and smarter reporting. The AI analyzer has demonstrated superior performance compared to human readers in identifying suspicious regions of interest (ROIs) in the lungs, pleura, and cardiac areas from chest radiographs. The use of AI aids increased the average sensitivity and area under the receiver operating characteristic curve (AUROC) by 14.2% and 4.4%, respectively. Another notable product, **Genki**, is an AI-powered tool developed by DeepTek for lung health screening, specifically designed to automatically screen digital chest X-rays for signs of tuberculosis (TB). Genki detects TB with an accuracy of 92%, sensitivity of 91%, specificity of 92%, and an AUROC of 96%. These tools utilize advanced AI and machine learning techniques to enhance diagnostic accuracy and efficiency in medical imaging.

## Patents

Display screen or portion thereof with graphical user interface, Patent number: D949895. Inventors: Amit Kharat, Aniruddha Pant, Ajit Patil, Vinit Mandhre, Viraj Kulkarni

## Commerfcialisation

Products- Augmento X- RAY-Chest X-Ray Reporting Automation Tool, Automated segregation of scans into low suspicion and high suspicion, Automated Structured Clinical Report Generation, Responsible AI (Post Deployment Surveillance). Genki is AI Powered Lung Health Screening Tool. The firm also offers Advanced AI Enabled Teleradiology Services with Quick Turnaround time, Structured & Quantified Reporting backed by Team of Expert Radiologists.

https://www.deeptek.ai/augmento-xray

# Autologous Stem Cell Treatment-Nirupa Vyas/Sushruta Innovations, Vadodara

Nirupa Vyas is the Managing Director of Sushruta Innovations and Wellness Pvt. Ltd., a company specializing in stem cell treatments for various medical conditions. Nirupa holds an MBA in Finance, an LLB, and a Ph.D. in Finance. Under her leadership, Sushruta Innovations received funding from the Kerala State Industrial Development Corporation (KSIDC) and is currently incubated at the Rajiv Gandhi Centre of Biotechnology in Kochi. Nirupa also serves as the Managing Director of Total Potential Cells Pvt. Ltd., a company focusing on stem cell research and therapies. Dr. Bhaskar Vyas, MS, a plastic surgeon with training from Edinburgh and a research fellowship at Hahnemann Medical College, Philadelphia, is the Chairman of the company. Dr. Bhaskar Vyas has been a pioneer in establishing the plastic surgery unit at the Medical College and M.S. University, Baroda, and has a long history of contributions to the field of regenerative medicine.

## Technology

Stem cell research dates back to the late 19th century when Theodor Boveri and Valentin Haecker coined the term "stem cell." The first clinical application of stem cells occurred in 1956 when French oncologist Georges Mathé performed a bone marrow transplant on five workers affected by a nuclear accident. The procedure was groundbreaking and paved the way for future stem cell therapies.

## Innovation

There are two main varieties of stem cells in the permissible area of research - Hematopoietic stem cells (HSCs) and mesenchymal stem cells (MSCs). Total potential Cells are specialised in MSCs. They have successfully isolated and cultured mesenchymal stem cells (MSCs) derived from bone-marrow, umbilical cord and adipose tissue that are multipotent stem cells. The properties of MSCs are honing to damaged tissues, tri-germinal translation, immuno-naive and immuno-protective.

Sushruta offers Autologous stem cell treatment for a range of orthopedics, neurological, and autoimmune conditions. Treatment developed for; Progressive Supranuclear Palsy, Optic nerve injury, Avascular Necrosis of Femur, Pulmonary Fibrosis, Chronic Kidney Failure, Diabetes I and II, Liver Cirrhosis, Muscular Dystrophy, Osteoarthritis of Knee, Diabetic Retinopathy, Multiple Sclerosis, Autism, Cerebral Palsy and Parkinson's Disease.

## Patents

PROCESS OF PREPARING MESENCHYMAL STEM CELLS FOR THE TREATMENT OF OSTEOARTHRITIS AS REGENERATIVE MEDICINE, Patent no 418407, Inventors- 1.Dr. Nirupa Vyas 2.Dr. Rajni Vyas

DIFFERENTIATION OF HUMAN ADIPOSE TISSUE DERIVED STEM CELLS TO ISLET CELL AGGREGATES (ICA) AND TREATMENT AS REGENERATIVE MEDICINE FOR DIABETES, Patent no 421156, Inventors- 1.Dr. Rajni Vyas 2.Dr. Nirupa Vyas

## Commercialization

Hospitals interested in collaborating can contact:

https://www.sushrutawellness.com/conditions.html

# Cell therapy for Eyes-Jogini Desai/Eyestem/ Bengaluru

Dr. Jogin Desai is the Founder and CEO of Eyestem, a pioneering company specializing in cell therapy for ocular diseases. Dr. Desai previously served as the CEO of Cenduit, the world's largest standalone randomization company, which had operations across Bangalore, Basel, Philadelphia, and North Carolina. He has brought together a team of experts to advance Eyestem's mission, including Dr. Rajani Battu, Co-founder and Chief Medical Officer, who is a senior consultant in Ophthalmology at Aster CMI Hospital in Bangalore and a fellow of the Royal College of Surgeons, Edinburgh. Dr. Rajarshi Pal, Co-founder and Chief Scientist, has extensive experience in regenerative medicine, having served as Assistant Professor at the Manipal Institute of Regenerative Medicine and worked at notable research institutions like the National Institute of Immunology and Stempeutics Research. Dr. Dhruv Sareen, Co-founder and Science Officer, is the founding director of the Induced Pluripotent Stem Cell (iPSC) Core at Cedars-Sinai BOG-RMI and has a Ph.D. from the University of Wisconsin at Madison. Eyestem has received investment from Endiya Partners.

## Technology

Macular Degeneration (AMD), particularly Age-related Macular Degeneration, is a leading cause of central vision loss in people aged 60 and above. There are two main types: dry (Atrophic) and wet (Exudative). Additionally, Inherited Retinal Diseases such as Retinitis Pigmentosa (RP) contribute significantly to irreversible blindness. There are currently no effective treatments for dry AMD, the more common form of the disease.

The Eyestem team is focusing on using Induced Pluripotent Stem Cells (iPSCs) to develop Retinal Pigment Epithelial (RPE) cells that could replace damaged cells in an allogeneic setting. Their proprietary product, Eyecyte-RPETM, uses a scalable protocol to generate authentic RPE cells in monolayer cultures. Another product, Eyecyte-PRPTM, aims to generate transplantable retinal photoreceptors from iPSCs to treat Retinitis Pigmentosa (RP).

## Innovation

Eyestem's innovative approach targets diseases like dry AMD, for which there are currently no available treatments. Their technology involves the use of iPSC-derived RPE cells to replace lost retinal cells. Eyecyte-RPETM represents a novel solution, utilizing a renewable source of cells to replace the damaged or lost RPE cells in patients with AMD. Additionally, Eyestem's Eyecyte-PRPTM leverages a patent-pending unified protocol to develop retinal photoreceptors that hold promise for replacing damaged photoreceptors in conditions like Retinitis Pigmentosa.

## Patents

Method for producing photoreceptor cells, retinal pigmented epithelial cells, and retinal organoids, Patent number: 11168301. Inventors: Rajarshi Pal, Rajani Battu, Reena Rathod, Harshini Surendran, Kapil Bharti, Deepak Lamba, Dhruv Sareen, Mahendra Rao, Sushma Nanjunda Swamy, Vijay Bhaskar Konala Reddy, Mohanapriya Rajamoorthy

## Commercialization

Lead product out of the platform is Eyecyte-RPETM aimed at treating Dry AMD (Age-related Macular Degeneration). Dry AMD is the largest cause of incurable blindness in people above 60 and affects roughly 170 million people worldwide. Pivotal efficacy and safety studies for this product have started at Oregon Health and Science University and startup expects to begin first in man studies for the same in 2024.

https://eyestem.com/our-product-pipeline/

# Custom Made Assistive Device-Arun Cherian/Rise Bionics, Bengaluru

Arun Cherian is the Founder of Rise Bionics, a company based in Bengaluru that specializes in developing affordable and high-performance prosthetic limbs. Cherian left his PhD program in Mechanical Engineering at Purdue University to focus on creating assistive devices. He previously conducted research at the University of California, Berkeley, on wearable exoskeletons designed to help paralyzed individuals walk. With a Master's degree in Mechanical Engineering from Columbia University and a Bachelor's degree from Anna University, Cherian has dedicated his career to improving the lives of amputees. In 2016, he was selected for the D-Lab Scale Ups Fellowship, which provided financial support, mentorship, and networking opportunities to help scale Rise Bionics.

## Technology

Rise Bionics leverages both traditional and innovative materials and methods to create its prosthetic devices. The company uses cane—a locally sourced, sustainable material—to build prosthetic legs that are lighter than similar devices on the market. This approach provides a dual benefit: it supports local artisans by creating a sustainable income stream and produces an affordable, lightweight prosthetic leg that enables users to walk, run, play, and even dance. The development of these prosthetics is influenced by open-source designs, like those from the University of Michigan and Shirley Ryan AbilityLab, which promote collaboration and advancement in the field of bionics https://www.opensourceleg.org/

## Innovation

The flagship product, Rise Legs, is a cane-based prosthetic that provides a cost-effective alternative to more expensive devices like carbon fiber prostheses. At a price point similar to the Jaipur foot, Rise Legs offer the advanced functionality of much more expensive devices. The product is up to four times lighter than similar prosthetics, making it easier for users to maneuver. The legs are assembled by local technicians and distributed in collaboration with organizations focused on aiding individuals with disabilities. Rise Bionics has also expanded its outreach by partnering with sports, dance, and martial arts schools to help amputees pursue their athletic and artistic dreams. Supported by the D-Lab Scale-Ups Fellowship, Rise Bionics is conducting a pilot clinical trial, setting up a clinic in Bangalore, and exploring potential expansion into the African market.

## Patents

https://www.frontiersin.org/journals/robotics-and-ai/articles/10.3389/frobt.2020.594196/full

## Commercialization

Rise Bionics has proven its technology on international platforms such as Cybathlon, where athletes using prosthetics compete in athletic events. Remarkably, athletes using Rise's $300 prosthetic leg outperformed competitors using prosthetics from major companies costing up to $100,000, winning two out of three races and setting the fastest times. Beyond prosthetics, Rise Bionics has developed a streamlined workflow to customize devices in hours rather than weeks. This process includes training paramedical professionals to use a handheld scanner for precise measurements, which are then used to create a custom mesh for the patient. https://risebionics.com/

https://amputeestore.com/blogs/amputee-life/rise-bionics-providing-high-quality-affordable-orthotics-and-prosthetics-in-india

# Digital Microscope with AI-Tathagato Rai Dastidar/SigTuple, Bengaluru

Tathagato Rai Dastidar is the Founder and CEO of SigTuple, a company that aims to revolutionize the healthcare domain using artificial intelligence (AI) and computer vision. A tech industry veteran with over 22 years of experience, Tathagato holds a BTech and Ph.D. in Computer Science from the Indian Institute of Technology, Kharagpur. He started his career in 2000 at National Semiconductor, where he developed software for semiconductor design automation. In 2015, he co-founded SigTuple with the vision to assist pathologists in their microscopy work, thereby enhancing efficiency and improving patient outcomes. Tathagato is credited with 11 US patents, six of which he is the primary author.

## Technology

SigTuple's innovation lies in its use of AI and digital microscopy to transform the traditional pathology workflow. Digital microscopy, specifically whole slide imaging (WSI), automates the process of capturing and reviewing microscopic images. This approach converts physical samples into digital "virtual slides," which can be reviewed by multiple experts remotely, enabling collaborative analysis and opening up possibilities for AI-based automated analysis. The core of this technology involves automated focusing processes that use machine learning algorithms to optimize the focus position of the sample under the microscope, enhancing the precision and accuracy of digital imaging.

## Innovation

SigTuple has developed an integrated hardware and AI-powered medical device that represents a significant breakthrough in digital microscopy. This device is the first AI-assisted digital pathology solution from India to receive clearance from the US FDA. It enables the digital imaging of physical samples through a microscopic lens, after which AI models analyze the images to extract and classify cells into over 30 different types. This technology allows pathologists to review samples remotely and significantly increases their efficiency by automating most of the review process. As a result, a single pathologist can handle a much larger volume of samples than was previously possible, reducing the need for additional manual reviews. SigTuple's method of applying deep learning for focus distance estimation in automated digital microscopy has shown superior generalization over multiple staining protocols, making it a robust tool for pathology labs.

## Patents

SigTuple's key innovation, the method, and system for auto-focusing a microscopic imaging system, is protected by US Patent number 11340439. The patented technology uses a machine-learned regression system, such as a Convolutional Neural Network (CNN), to receive and analyze multiple images of a sample captured at different focus positions. By calculating the difference between these images in terms of pixel values, the system determines the optimal focus position.

## Commercialization

igTuple's flagship product, the AI 100, is an in-vitro diagnostic device designed to automate manual microscopy in diagnostic laboratories. This device uses robotics and AI to digitize any biological sample on a glass slide, facilitating AI-assisted remote review. SigTuple has also developed specialized AI applications such as "Shanti," which analyzes blood cell morphology and pre-classifies white blood cells (WBCs), red blood cells (RBCs), and platelets in peripheral blood smears, and "Shrava," which analyzes urine sediment and pre-classifies multiple elements present in the sediment.

https://www.sigtuple.com/

# Green Crematorium-Arjunbhai Mohanbhai Paghdar/SwargaRohan, Junagadh, Gujarat

Arjunbhai Mohanbhai Paghdar, Grassroots Innovator of Gujarat developed a more efficient & socially acceptable wood based cremation system. He was awarded at NIF's 8th National Grassroots Innovation Awards for his brick/block making machine. He also made a wood splitting machine, cow dung log and pot making machine and a mobile chabutra (bird feeding platform).

## Technology

There have been many attempts to develop green cremation systems acceptable to the family of the diseased. IIT-Ropar has developed a technology that involves smokeless cremation of the body. This concept is based on a wick stove, in which the wick glows yellow when lighted and is converted into smokeless blue flame through an air combustion system. The cart has wheels and can be transported without a lot of effort. There are stainless steel trays on both sides for easy ash removal.

## Innovation

Shri. Arjunbhai Mohanbhai Paghdar, Grassroots Innovator of Gujarat developed a more efficient & socially acceptable wood based cremation system. The system is designed with an innovative approach to minimize fuel wood consumption and reduce air and water pollution. Its ritual enabling design makes it a user acceptable alternative amongst the various available methods. SwargaRohan Cremation Furnace is capable of reducing fuel wood consumption by over 70 to 75%. Generally, cremation system a human body may consume around 400-500 kg wood. Swargaroahan consumes less than 100 kg of wood and abiding with all the faiths of religion. This cremation system saves time.

Arjunbhai has designed a biomass gasification based cremation process, wherein a closed mummy shape is made of refractory bricks to ensure minimum heat loss. He has placed doors at front and rear end of structure to perform Hindu rituals. The inner side of the top cover is filled with cera-wool, which can tolerate high temperature. Blowers and nozzles are provided which release air to accelerate the cremation process. Charcoal filters and caustic soda filters have been provided for air filtration and a chimney for air exhaust.

https://nif.org.in/innovation/modified-wood-based-culturally-accepted-crematorium/1085

## Patents

https://grid.undp.org.in/practices/9o3j0lGA4zaFW6#6/21.289/70.169

## Commercialization

The average time taken to burn a dead body (about 80kg weight) is 70-90 minutes while consuming 70 to 80 kg of wood as compared to 3-4 hours in conventional method consuming around 400 kg of wood. The cost of modified crematorium is about Rs. 2 lakh.

https://swargarohan.info/

# Iron Oxide Transdermal Patch to Treat Anemia-Dr. Avinash R. Tekade, Pune

Dr. Avinash R. Tekade is a Professor and Head of the Department of Pharmaceutics and Pharmaceutical Quality Assurance at Marathwada Mitra Mandal's College of Pharmacy in Pune. With over 23 years of experience in education and drug delivery research, Dr. Tekade specializes in areas such as nanotechnology-based drug delivery systems, pharmacology, and technology transfer. He holds a Master's degree in Pharmaceutical Sciences from Birla Institute of Technology and Science, Pilani, and a Ph.D. from North Maharashtra University, Jalgaon. Dr. Tekade has published 53 papers and holds 5 patents, with 2 patents already granted. His contributions to science and technology were recognized with the Prof. Man Mohan Sharma Science and Technology Award in 2019.

## Technology

Iron deficiency is the predominant cause of anemia across countries and in both sexes, with women more commonly afflicted. women and women from underdeveloped and developing countries. The currently available formulations are in the form of tablets, liquid orals and IV injections. The absorption of iron from oral formulations is erratic and causes gastric disturbances in around 10% of patients whereas the intravenous administration is invasive method and need medical expert. Moreover, the chances of missing the doses are more in working women's and thus need such a drug delivery system which is when administered once can deliver the drug for prolonged period and thereby may increase the patient compliance.

## Innovation

Prof. Avinash team has made a breakthrough by innovating alternative route for delivery of iron oxide to treat mild to moderate form of anaemia by formulating patient friendly transdermal iron patch using natural permeation enhancer for increasing its permeation through skin to replenish the iron loss from the body. This transdermal patch maintains iron level in body without side effects that are associated with oral dosage forms so as to improve the quality of life of the patient and to make the target group healthy. Salient Features of Transdermal Patch; Prepared transdermal patches contain Iron oxide solid lipid nanoparticles for increasing the permeation through skin using natural permeation enhancer and therefore may replenish the iron loss from the body. Formulated patient friendly (Convenience of application, Nomore gastric disturbances, No skin irritation) and economical transdermal patches help maintain the iron concentration in the blood particularly in women. Proof of Concept/ Outcome of Preclinical Studies in Rabbits.

## Patents

List of patents filed by the University:

https://mmcop.edu.in/ipr-cell/

A stable pharmaceutical composition for transdermal drug delivery, patent no 202121000493, innovators- Dr. Manohar J. Patil, Mr. Shekhar Holkar, Dr. Avinash Tekade.

## Commercialization

Patent available for licensing. Contact- Dr. Avinash R. Tekade, Professor & Head, Department of Pharmaceutics, Marathwada Mitra Mandal's College of Pharmacy, PCNTDA, Thergaon (Kalewadi), Pune-411033 (M.S.) Contact: 9371152536 Email:avitekade@gmail.com

# Kidney testing at home-`Sumona Karjee Mishra/Prantae Solutions, Bhubaneswar

Dr. Sumona Karjee Mishra is the Founder and CMD of Prantae Solutions, a pioneering health-tech startup based in Bhubaneswar. With a strong academic foundation in microbiology from Delhi University and a Ph.D. from the International Centre for Genetic Engineering and Biotechnology, Dr. Mishra ventured into entrepreneurship after winning the first-ever Young Entrepreneurs Scheme co-sponsored by DBT, India, and BBSRC, UK. Her co-founder, Dr. Aseem Mishra, holds a Ph.D. in Biotechnology from JNU and serves as an Associate Professor at KIIT University School of Biotechnology. Prantae Solutions has garnered funding from micro, small, and medium enterprises, as well as various grants from government and non-profit organizations, including DST-NIDHI-PRAYAS, DBT-BIRAC BIG, BPCL Ankur, and TATA Trust-Harvard SAI Livelihood Creation.

## Technology

The kidneys play a crucial role in filtering the blood and preventing the loss of vital proteins such as albumin in the urine. When kidney function declines, small amounts of albumin leak into the urine—a condition known as microalbuminuria. Similarly, creatinine, a waste product from muscle metabolism, is filtered out of the blood by the kidneys. A decreased creatinine level in the blood indicates reduced kidney function. Traditional methods for detecting creatinine levels, such as colorimetric, spectrophotometric, and chromatographic techniques, are time-consuming and not suitable for point-of-care use. Creatinine nanomaterial-based biosensors, on the other hand, provide a rapid and sensitive alternative, capable of detecting a wide range of creatinine concentrations (0.004-30,000 $\mu$M) with detection limits between 0.01 $\mu$M and 520 $\mu$M.

## Innovation

Prantae Solutions has developed the world's first clinical-grade, accurate point-of-care device for estimating the Albumin to Creatinine Ratio (ACR) and Urine Albumin to Creatinine Ratio (UACR)—key parameters for assessing kidney health as per KDIGO, the National Kidney Foundation, and the American Diabetes Association guidelines. Their flagship product, Proflo-U®, is a comprehensive kidney health monitoring platform. It includes an ergonomic, sealed vial with formulated reagents for detecting kidney biomarkers in urine, which transforms into a cuvette for precise measurements upon the addition of urine. The Proflo-U® urine albumin test uses a nanosensor for rapid and accurate estimation of protein levels (30-1000 mg/L).

## Patents

A relevant patent titled "DEVICE AND METHOD FOR DETECTING CREATININE AND ALBUMIN TO CREATININE RATIO" (Publication number: 20180149612) has been filed by inventors Vinay Kumar, Navakanta Bhat, and Nikhila Kashyap Dhanvantari Madhuresh, and is assigned to IISc, Bangalore. This invention pertains to an electrochemically active device capable of detecting and measuring creatinine and albumin levels in biological samples, simultaneously determining the ACR. The point-of-care biosensor described in this patent facilitates the quantitative measurement of these biomarkers in urine and blood samples, providing a rapid and reliable tool for clinical use.

## Commercialization

The device measures albumin, creatinine, and ACR, ensuring precise results for informed clinical decisions and effective treatments. Delivers rapid results in less than 2 minutes, enabling timely clinical decisions and swift patient care. Enabling seamless and wireless data transfer. It is portable and compact, making it easy to transport between locations and use in a variety of settings. Proflo-U harnesses the power of advanced AI algorithms.

https://www.proflou.com/products

# Multi Disease Prognosis Provider-Sanjay Jeswani/Perkanttech, Vizag

Saniya Jeswani is Co-Founder & CEO, Perkant Tech | Top 75 Women Entrepreneurs. A graduate from Symbiosis University of Applied Science, directed Perkant Tech's ascent on numerous international platforms including World Health Organisation (WHO) HQ - Geneva, GIZ (German Development Agency), and various global embassies, cultivating pivotal partnerships. Lokant Jain is Co-Founder and CTO at Perkant Tech, a graduate from IET DAVV, Indore represents startups on the medical device committee of FICCI. The startup received grants from AIM, Niti Aoyog.

## Technology

A photoplethysmograph (PPG) is a simple medical device for monitoring blood flow and transportation of substances in the blood. It consists of a light source and a photodetector for measuring transmitted and reflected light signals. Clinically, PPGs are used to monitor the pulse rate, oxygen saturation, blood pressure, and blood vessel stiffness.

## Innovation

Abhay Parimiti, the startup flagship product, harnesses a blend of advanced photoplethysmography (PPG) sensors and artificial intelligence (AI) to deliver a comprehensive and accessible health screening solution. This core technology operates on the scientific principle of detecting volumetric changes in blood flow through light-based sensors, which are then analyzed to assess various health parameters such as heart rate, oxygen saturation, and blood pressure. PPG technology is the backbone of the device, enabling non-invasive monitoring by emitting light into the skin and measuring the light absorption changes caused by blood flow dynamics. This method provides real-time data on blood volume changes, which are crucial for assessing cardiovascular health. device achieves clinical-grade precision through individual calibration and a high sampling rate of 3000 samples per second (achieved by our proprietary algorithms), tailored to diverse patient demographics. With an added layer of various calibration modes through gold-standard systems periodically (ex: A sphygmomanometer) we ensure high accuracy even with a finger based measurement methods. The data captured by PPG sensors are processed using over 100 proprietary algorithms developed in-house. These algorithms enable the device to interpret complex biometric data, turning raw signals into actionable health insights.

## Patents

A SYSTEM COMPRISING OF AN APPARATUS FOR MEASURING AND ACQUIRING DISTRESS DATA FROM THE HUMAN BODY, 20230395210, Inventors: Lokant JAIN, Saniya JESWANI

## Commercialization

Andhra Pradesh MedTech Zone (AMTZ) is the medical technology park with Common Manufacturing Facilities & Common Scientific Facilities located in Nadupuru village area of Visakhapatnam. Centre for MedTech Innovation and Rapid prototyping facility is operated by M/s T3D labs Private Ltd (Think 3D) and is primarily intended for medical device development and customized implants.

https://perkanttech.com/#our-product

# Rural Surgery Innovations, Ganaraj/RSI/ Nagaland

Dr. Gnanaraj is a urologist trained at Christian Medical College, Vellore, who has dedicated his career to enhancing surgical access in rural settings. He has worked in several rural hospitals across India with organizations like Friends Missionary Prayer Band and Emmanuel Hospitals Associations. His experience in these regions underscored the need for innovative, low-cost surgical solutions tailored to the unique challenges of rural healthcare environments. Collaborating with international universities and local Indian medical equipment manufacturers, Dr. Gnanaraj has developed several groundbreaking technologies to make surgeries more accessible and affordable in rural areas. Mr. Kesochie Chakhesang, with a background in media, communications, and economics from Nagaland, India, plays a vital role in managing and organizing the services of Rural Surgery Innovations (RSI).

## Technology

RSI's key innovation, the **gasless laparoscope**, allows laparoscopic surgeries without gas insufflation and general anesthesia, making them accessible in low-resource settings. This device lifts the abdominal wall mechanically to create an operative space, reducing infrastructure needs and enabling quicker patient recovery. An improved version, **RAIS** (Retractor for Abdominal Insufflation-Less Surgery), was developed with the University of Leeds.

## Innovation

RSI also developed the **Laptop Cystoscope**, which simplifies cystoscopy by using a laptop's USB port, making it ideal for rural areas lacking specialized equipment. Their **Gas Insufflation Less Laparoscopic Surgeries (GILLS)** technique allows laparoscopic procedures using spinal anesthesia, further reducing costs and infrastructure requirements. Additionally, RSI has innovated methods for removing renal stones through the urinary passage, providing a less invasive option for rural patients.

## Patents:

Paper- https://academic.oup.com/inthealth/article/8/6/367/2433262

SYSTEMS, DEVICES, AND METHODS FOR PREVENTING OR REDUCING LOSS OF INSUFFLATION DURING A LAPAROSCOPIC SURGICAL PROCEDURE, Publication number: 20210401462, Inventors: Adam R. Dunki-Jacobs (Cincinnati, OH), Mark S. Ortiz (Milford, OH), Jonathan R. Thompson (Cincinnati, OH), Richard P. Nuchols (Williamsburg, OH), Caleb J. Hayward (Goshen, OH), Robert T. Means, III (Cincinnati, OH), Saylan J. Lukas (Cincinnati, OH)

## Commercialization

Rural Surgery Innovations Private Limited (RSI) is a company formed in association with the International Federation of Rural Surgeons (IFRS) and the Association of Rural Surgeons of India (ARSI) to assist in carrying out rural surgery specific research and development of the products of such research.

The RSI Base Hospital is in the "Chellappa Nursing Home" in Tiruchirappalli, Tamil Nadu in India. It is a 15 – bed surgical facility with capabilities for open, laparoscopic, and endoscopic surgeries. It was started by Dr. JDJ Chellappa one of the senior surgeons in Tiruchirappalli three decades ago. Dr. Gnanaraj Jesudian is the surgeon at the base hospital and Dr. JDJ Chellappa takes care of the out patients. The main office of RSI is in Nagaland, in Northeast India where the majority of demand and services for training and equipment currently happens. This is where Mr. Kesochie works from and organises the services of RSI. Dr. Gnanaraj has spent a significant portion of his professional career working in this part of India and is familiar with the needs of the patients here.

https://www.ruralsurgery.in/products

# Tin Coated Coronary Stent System-SCTIMST, Thiruvananthapuram

Mr. Muraleedharan C. V. is Scientist 'G' Sree Chitra Tirunal Institute for Medical Sciences & Technology, Trivandrum. (https://www.sctimst.ac.in/people/muralicv). He is Co-PI for this project. Team has Subhash NN (PI) (https://www.sctimst.ac.in/People/subhashnn) and Dr. Harikrishnan S (PCI) (https://sctimst.irins.org/profile/289565),

Significant contribution have come from Dr. Sujesh S, Dr. Sanjay G, Dr. Umasankar, Dr. Sachin, Dr. Sabareeswaran, Mr. Ramesh Babu, Mr. Rajeev A, Mr. Subhash Kumar, Mr. Aneesh, Mr. Polson, Mr. Binu & Various Divisions & Departments of SCTIMST. Selected among top 3 innovations in health by AEGIS GRAHAMBELL AWARD, and top 3 innovations in BOEING BUILD 3.0

## Technology

Puel and Sigwart discovered the first coronary stent (in 1986), which was able to prevent acute vessel closure and late constrictive recoil. Coated stents are new generation stents coated with a thin polymer membrane for vascular therapy. Coated stents increase tissue granulation and suppress thrombosis compared to bare-metal stents (BMS). A proprietary process was developed to coat titanium-nitride-oxide on the stent surface, based on plasma technology that uses the nano-synthesis of gas and metal. Preclinical in vitro and in vivo investigation confirmed blood compatibility of titanium (nitride-) oxide films. Titanium-nitride-oxide-coated stents demonstrated a better angiographic outcome, compared with bare-metal stents. Stents coated with titanium-nitride-oxide demonstrated biocompatibility in preclinical studies: they inhibit platelet and fibrin deposition, and reduce neointimal growth.

In observational and non-randomized studies, titanium-nitride-oxide-coated stents were associated with adequate safety and efficacy outcome. In randomized trials of patients with acute coronary syndrome, titanium-nitride-oxide-coated stents were associated with a better safety outcome, compared with drug-eluting stents; efficacy outcome was comparable.

## Innovation

The Chitra coronary stent consists of the following components:

1. L605 based Metal Stent platform

2. Titanium Nitride Ceramic coating

Design and choice of materials completed. 12 TiN coated coronary stent are prototyped. The safety and performance of the design has been assessed in vitro and in silico.

## Patents

Patent granted, 357096 dated 29/01/2021

List of Foreign patents filed- https://www.sctimst.ac.in/Academic%20and%20Research/Research/Intellectual%20Property%20Rights/resources/InternationalPatents_21.05.2024.pdf

## Commercialization

EOI advertised:

https://sctimst.ac.in/technology-transfer/TiN%20coated%20coronary%20stent%20system/

FLOW DIVERTER STENT – Technology transferred to M/s BIORAD MEDISYS Pvt. Ltd, Pune.

AORTIC STENT GRAFT (Straight for thoracic portion of the aorta.) - EOI invited.

ASD OCCLUDER – Technology transferred to M/s BIORAD MEDISYS Pvt. Ltd, Pune.

# Virtual Autopsy-Hemant Naik, Virtual Autopsy India, Bangalore

 Dr. Hemanth Naik is Consultant PMCT(Virtual Autopsy) and COO Virtual Autopsy India. A seasoned forensic specialist with graduation from KMM Medical College and PG in Forensic Medicine from Bangalore Medical College and Research Institute. He is Chief Medical Officer at Virtual Autopsy Global Solutions, UK. He is associated with Associated with iGene London Ltd working on ` Forensic Information System for supporting workflow management to iDASS (Digital Autopsy) and integration with PACS and HL7 compliant information systems'.

Ash Govind is Founder & CEO at Virtual Autopsy Global Solutions™. UK. A graduate from University of Leicester with BA (Hons) in Humanities with over 35 years of experience in the global Death-care sector, with over a decade exclusively in the forensics and post-mortem imaging and visualisation arena. To complement his work in Virtual Autopsy, Ash is also engaged in Mass Fatality recovery and Disaster Management response and is a certified Member of the Institute of Consulting (MIC) and a Member of the Chartered Management Institute (MCMI).

## Technology

An autopsy (postmortem examination, autopsia cadaverum, or obduction) is a highly specialized surgical procedure that consists of a thorough examination of a corpse to determine the cause and manner of death and to evaluate any disease or injury that may be present. Virtopsy is a word combining 'virtual' and 'autopsy' and employs imaging methods that are also used in clinical medicine such as computed tomography (CT), magnetic resonance imaging (MRI), etc., for the purpose of autopsy and to find the cause of the death. Virtopsy can be employed as an alternative to standard autopsies for broad and systemic examination of the whole body as it is less time consuming, aids better diagnosis, and renders respect to religious sentiments.

Dr. Michael Thali, a professor at the University of Zurich and co-founder of The Virtopsy Project, introduced virtual autopsies in 1999, creating permanent 3D models of bodies that can be easily accessed and shared for second opinions. This technique has become common practice in Swiss forensic investigations.

## Innovation

VizTouch3D™ forensic specific touchscreen Autopsy Tables provide Pathologists with a natural user interface through which to conduct non-invasive autopsies. Height adjustable and with electric tilt for vertical and horizontal use (specific to the model), the multi-touch solution comes with a wireless remote control and heavy-duty wheels for ease of manoeuvring (specific to the model). We manufacture 3 models in multiple screen sizes, complete with High-Performance Computing (HPC), Viz3D™ Software, and pre-installed real dead body data sets in multiple modes of death.

## Patents

Three-dimensional human body virtual autopsy table, https://eureka. patsnap.com/patent-CN107689186A

## Commercialization

All India Institute of Medical Sciences (AIIMS)-Mangalagiri has equipped itself with a 'virtual dissection laboratory', a synchronised multiple visualiser system. The system consists of Anatomage convertible table, Anatomage wall with 4 synchronous screens, and RealSim tilt table with fully pre-loaded complete 3D anatomy of real human cadavers customisable from user imported CT/MRI data compatible with upcoming PACS for radiological review and virtual autopsy paradigm.

https://virtual-autopsy.com/viztouch-3d-dissection-table/

# UltraHand: Robotic Hand Rehabilitation Device-Swapnil Ajit Bukshete/Queliz Lifetech, Pune

Swapnil Bukshete is a graduate from Savitribai Phule Pune University in mechanical engineering followed by PG Diploma in Additive manufacturing from COEP Technological University. He worked as Research Assistant at BETiC, IITB and was BIRAC SPARSH Fellow at IIT Knapur. The innovation received grants such as NIDHI PRAYAS and BIRAC BIG. Dr. Sandeep Anasane, Assistant Professor, Co-PI, BETiC-COEP Pune is their technical mentor. Dr. Sandeep is Head, Dept. of Manufacturing Engineering & Industrial Management, COEP Technological University, Pune (Formerly College of Engineering Pune- COEP) and Member, National Steering Committee, AICTE-IDEA lab

(Nominated by P.S.Kannan, General Manager – KonguTBI Technology Business Incubator @ Kongu Engineering College (TBI@KEC))

## Technology

In 80% of injuries such as burns, strokes, arthritis, cerebral palsy, surgical interventions hand is involved, many patients suffer from improper rehabilitation due to the lack of accessible and affordable devices. Hand rehabilitation devices (HRDs) are devices designed to provide the hand with passive, active, and active-assisted rehabilitation therapy. Rehabilitation therapy is defined by three main forms: passive, active, and active-assisted therapy. Passive rehabilitation therapy is when the only force applied for the patient to complete their exercise is external, meaning the patient inputs no effort. This allows patients with more limited functions to begin their rehabilitation therapy. Active therapy is when no external force is used to complete an exercise, and patients need to exert all their effort to complete the exercise. This step is crucial for the patient, as active exercises strengthen the hand muscles and solidify their neuroplasticity. Active-assisted therapy combines both forms, providing the patient with an opportunity to strengthen their muscles much more efficiently. Rehabilitation requires all three phases to return the patient to proper motor function.

## Innovation

UltraHand is a revolutionary robotic hand rehabilitation device that offers both active and passive rehabilitation for comprehensive hand mobilization. Its unique selling proposition lies in its affordability and advanced functionality, making it accessible to a wider population. UltraHand ensures complete movement of MCP, PIP, and DIP joints, crucial for restoring hand functionality. It features customizable rehabilitation modes significantly enhancing the rehabilitation process and ensuring faster recovery. Its lightweight, portable design allows for immediate deployment, making it a vital tool for effective hand rehabilitation. Key Features- Mode of Rehabilitation, Passive Rehabilitation: Continuous Passive movement device prevent stiffness. Active Rehabilitation: Helps to strengthen the muscle, Complete Hand Mobilization,0 to 90° Metacarpophalangeal (MCP) Joint Movement, 0 to 110° Proximal Interphalangeal (PIP) Joint Movement

0 to 70° Distal Interphalangeal (DIP) Joint Movement, Thumb Movement

## Patents

Indian Patent

Title: "A ROBOTIC HAND REHABILITATION DEVICE AND METHOD THEREOF"

Application Number: 202221046107

Patent No: E-5/2070/2023/MUM

Dated: 12/08/2023

The patents were assigned to College of Engineering, Pune.

## Commercialization

Not Commercialized. Under Clinical Study Phase

**https://queliz.in/product**

## Profiled in 2022

## GROUP B- HEALTHCARE

1. Assistive Exoskeleton Devices – GenElek

2. Chimeric Antigen Receptor T cell (CAR T) therapy: Immunoadoptive

3. Digital Stethoscope: Ayu Devices

4. Genomic test for TB patients: Haystack Analytics

5. RT-PCR kit 'Om' for Omicron variant- CDRI

6. Bi Level Positive Airway Pressure (BiPAP) System Portable Ventilator- NAL

7. Asthma Control- Briota Technologies

8. Vein Tracking - MEDTRA INNOVATIVE TECHNOLOGIES

9. Portable Pathology Lab and La-Bike- Accuster Technologies

10. Non-Invasive Gluco meter- Vivalyf Innovations

11. Cancer Margin Detection- TeraLumen Solutions

12. Wearable Solutions for Chronic Knee Pain & Spine-ASHVA

13. Suitcase OCT- Tishyas Medical Device Development Solutions

14. Early Diagnosis of CKD- Prantae Solutions (OPC)

15. Robotics Surgical Intervention Platform- Comofi Medtech

16. 3D Bio-Printer- Avay Biosciences

17. Smart Posture Trainer-Neukelp Innovation Technology

18. Diagnostic Device for Malaria Chikungunya & Dengue- Ameliorate Biotech

19. Foetal Genitals- Purplas IT Services

20. Cardiac Arrest- Inofinity Research and Development

21. Wound Healing with Biomaterials- Alicorn Medical

22. Wound Healing through Silk proteins- fibroheal

## Profiled in 2023

### GROUP C: HEALTHCARE

1. AI based Breast Cancer detection- NIRAMAI Health Analytix, Bangalore

2. COVIHOME RNA test kit- Prof. Shiv Govind Singh, IITH,

3. DNA Clean-up kit- Procyto Labs Pvt. Ltd, Bhubaneswar, Odisha,

4. Dialysis-grade membranes - Prof. Jayesh Bellare, IITB

5. Germicidal fabric technology - Praveen Kumar Vemula Lab, inStem, Bangalore

6. Insect Venom derived Anti-aging Peptide for Dermal Application- Prof. S Ramaswamy Lab, inStem, Bangalore

7. Jaipur Belt- Exoskeleton- Newndra Innovations, Jaipur

8. Lab-on-chip, Microfluidic chips- Prof. Anil Prabhakar, IITM

9. Live cell imaging- Fluorescent Bioprobes Pvt Ltd, Goa

10. mRNA Covid-19 vaccine- Gennova Biopharmaceuticals Limited, Pune

11. Magnetic Resonance Imaging (MRI) Scanner- Voxelgrids Innovations Pvt Ltd, Bangalore

12. Microneedles for painless injections- Suman Pahal, inStem, Bangalore

13. Phototherapy device for neonatal Jaundice- Heamc Healthcare, Hyderabad

14. Portable inverted microscope - Prof. Debjani Paul, IIT Bombay

15. Protonated Bio-adhesive Polymer technology (PBT-Axio Biosolutions, Bangalore

16. Poorti, Post-Mastectomy Kit- Aarna Biomedical Products, Faridabad

17. Robotic Surgery- SS Innovations, Gurugram

18. Swasa- Respiratory Healthcare- Salcit Technologies, Hyderabad

19. ToucHb- A Non-invasive Aaemia Screener - Biosense Technologies Pvt. Ltd., Bhiwandi, Maharashtra

20. Video Laryngoscope- Dr. Kumaresh Krishnamoorthy, Bangalore

21. Visual medical data with AI- SigTuple Technologies Pvt. Ltd, Bangalore

22. Voice Prosthesis- inaumation medical devices, Bangalore

23. Wearable EEG device- Neuphony, Noida

# Water & Environment (2024)

# Bio Degradable Pen-Saurabh Mehta/Note, New Delhi

Saurabh (H.) Mehta is Entrepreneur in Sustainable Solutions. Saurabh Mehta, an engineer from BVCOE Delhi and an ICICI Fellow in Rural Development, has a diverse background in promoting sustainable and innovative solutions. He has undertaken various projects, including setting up solar-powered night schools in Bihar with the Aga Khan Rural Support Program, creating the "Storyteller" initiative at Agents of Nature to inspire young minds, and founding BioQ, a corporate gift shop that introduced plantable stationery. Currently, he is focused on his startup, NOTE (No Offense To Earth), which has received support from the Startup India Seed Fund Scheme.

## Technology

Plastic pens are a significant environmental concern, with over 50 billion discarded annually. These pens are not typically recycled and can take hundreds of years to decompose, contributing to long-term environmental pollution. While there are recycling programs for writing instruments, such as the one created by Bic and TerraCycle, the challenge remains significant due to the widespread use and disposal of plastic pens.

## Innovation

Saurabh Mehta and his team developed a fully biodegradable ballpoint pen to address this issue. This pen uses a recycled paper refill, non-toxic ink, and materials like metal, paper, or bamboo for the body. The core innovation lies in creating a plastic-free refill. After experimenting with various materials for over four years, they developed a recycled paper cylinder with a precise 2mm internal diameter. A breakthrough was achieved with a vegetable oil-based coating that prevents ink leakage and absorption, providing an industry-standard 18-month shelf life.

## Patents

1. **A Biodegradable Writing Device and a Method to Manufacture the Same** *Patent Number:* WO2023079356A1 WIPO (PCT) *Inventor:* Venkata RamaKrishna Sastry Dwibhashyam This patent covers the method for manufacturing a biodegradable writing device, detailing the steps of creating a pen barrel from recycled paper, applying a water-repellent coating, and assembling the components to produce a fully biodegradable pen.

2. **Completely Biodegradable Refill Free Pen** *Patent Number:* WO2021260739A1 WIPO (PCT) *Inventors:* Navneet Kaur, Anjali Anjali, Shivam Aggarwal This patent describes an entirely eco-friendly, 100% biodegradable pen free from harmful plastics, metals, or chemicals. It includes innovations such as a wooden nib, a paper mache body with fly ash for durability, and a coating that allows for easy decomposition and recycling in the soil..

## Commercialization

https://thenote.earth/shop/

# Building Houses with Plastic Waste-Prashant Lingam/Bamboo House India, Hyderabad

Prakash Lingam and Aruna lingam founded Bamboo House, a social enterprise. Bamboo House India, go beyond traditional approaches to create sustainable and affordable livelihood models using bamboo and plastic waste. The work encompasses green and recycled furniture, interiors, and handicrafts, housing, crafts etc not only addressing environmental challenges but also generating meaningful employment opportunities for rural and tribal communities. This socially driven initiatives have resonated globally, earning recognition and features in over 1500 local and international media platforms such as Brut, Entrepreneur, BBC, World Economic Forum, French, Russian, Australian media to name a few. The innovative approaches have been studied and praised by leading academic institutions, including Harvard, Cornell, and ISB. The Social Entrepreneurs received over 20 entrepreneurial awards.

## Technology

The EcoARK Pavilion in Taipei, Taiwan, is an excellent example of an amazing house built from recycled plastic. National Geographic describes the EcoARK Pavilion as the benchmark for the future of green buildings due to its magnificent and stylistic design. The EcoARK Pavilion was built in 2010 using 1.5 million recycled plastic bottles. It's a 9-story exhibition hall with a floor area of almost seven basketball courts. Architect Arthur Huang designed the pavilion to withstand the forces of nature, like earthquakes and fires. The hall is designed with hollow polli-bricks made from recycled polyethylene terephthalate (PET) bottles. The bricks were designed with interlocking grooves to fit together like LEGOs. Therefore, they need only a little sealant for attachment. The magnificent pavilion can accommodate hundreds of visitors and is an incredible example of what can be built with recycled plastic.

La Casa de Botellas is a plastic house built by Alfredo Santa Cruz in Puerto Iguazu, Argentina. Alfredo built this house to promote social and ecological responsibility. It was built using 1,200 PET plastic bottles and 1,300 tetra-pack cartons.

The Plastic Bottle Village is a community of houses built by Robert Bezeau out of recycled plastic. The village is located on the island of Bocas del Toro in Panama. Robert collected more than one million plastic bottles for use in this project. This served as a path to eco-friendly building in Panama. He filled the plastics with sand for insulation and structure before using them to build walls.

A waste collector in Karnataka constructed a home for himself out of recycled plastic waste while collaborating with Plastics For Change India Foundation.

## Innovation

Social enterprise specializing in constructing low-cost shelters using bamboo, agricultural and plastic waste and provide livelihood opportunities to rural, tribal, and waste picking communities at the base of the economic pyramid. The organisation also specialises in designing social business models with scrap tyres, discarded drums, textile waste and with other waste and eco materials and creating green livelihoods in the process. Discarded plastic bottles are first shredded and then melted into PET foam, a core material used by the composite industry. This is then converted into panels to create the exterior of the house.

## Patents

Open Source Waste Plastic Granulator, https://pure.psu.edu/en/publications/open-source-waste-plastic-granulator

## Commercialization

The first house Lingam built in Hyderabad's Uppal was of about 800 square feet. It was constructed using 7 tonnes of plastic.

https://www.bamboohouseindia.org/recycled-plastic-house

# Carbon Quantom Dots-Dr. Binoy K Saikia/ NEIST, Jorhat

Dr. Binoy Kumar Saikia, a Principal Scientist at the North East Institute of Science & Technology (CSIR-NEIST) in Jorhat, has pioneered the development of fluorescent carbon quantum dots (CQDs) from Indian coal. His work aligns with the "Atmanirbhar Bharat" initiative by providing an indigenous, patented CQD technology that offers a range of applications and contributes to import substitution. He was honored with the prestigious Shanti Swarup Bhatnagar Award for his contributions to science and technology.

## Technology

Coal, traditionally considered a non-renewable energy source, is being explored as a feedstock for carbon-based materials. Among these, carbon quantum dots (CQDs) have garnered significant attention due to their small size (less than 10 nm), low toxicity, robustness, optical and chemical stability, and ease of fabrication. CQDs are a type of carbon nanomaterial that also includes other forms like carbon nanotubes, fullerenes, graphene nanosheets, and more. These materials have diverse applications in bio-imaging, cell imaging, sensing, photovoltaic devices, and catalysis, owing to their unique optical and electronic properties.

## Innovation

Dr. Saikia's innovation involves a method to produce blue-fluorescence emitting carbon dots from sub-bituminous tertiary high sulfur Indian coals. The process utilizes an ultrasonic-assisted wet-chemical method to fabricate size-controlled carbon dots in a simple and environmentally friendly manner. This approach leverages low-quality Indian coals, including coal washery rejects (waste), to create valuable nanomaterials that can be used for detecting heavy metal ions like $Hg^{2+}$ and $Cr^{6+}$ in water.

## Patents

**Paper**: A. Boruah and B. K. Saikia, "Chemical Fabrication of Efficient Blue-luminescent Carbon Quantum Dots from Coal Washery Rejects (Waste) for Detection of $Hg^{2+}$ and $Cr^{6+}$ Ions in Water," *ChemistrySelect*, 2022, 7, e202104567.

**Patent**: "A Process for the Preparation of Blue-Fluorescence Emitting Carbon Dots (CQDs) from Sub-bituminous Tertiary High Sulfur Indian Coal," Patent number: 010655061. Inventors: Binoy Kumar Saikia, Tonkeswar Das, Sonali Roy, Bardwi Narzary, Hari Prasanna Deka Boruah, Manobjyoti Bordoloi, Jiumoni Lahkar, Dipankar Neog, Danaboyina Ramaiah., DIPANKAR NEOG, DANABOYINA RAMAIAH.

## Commercialization

Applications for Quantum Dots- Biological imaging and diagnostics: stem cells and cells tracking, fluorescent labeling of cellular proteins, pathogen and toxin detection, etc.In vivo animal imaging, Drug delivery, Photodynamic therapy

Light indicators for biological coding, Biosensors, Photovoltaic devices

Solar cells, Light emitting devices, Catalysis.

The global Carbon Quantum Dots (CQD) market was valued at USD million in 2022 and is anticipated to reach USD million by 2029, witnessing a CAGR during the forecast period 2023-2029. The key global companies of Carbon Quantum Dots (CQD) include Jiangsu Xianfeng Nano, Jintanchi, Beijing Beida Jubang, Nanjing NanoJanus, Suzhou Xingshuo Nanotech, Sigma-Aldrich, American Elements and ACS MATERIAL, etc.

Industry partners can contact:

Coal and Energy Research Group

CSIR North East Institute of Science And Technology

CSIR – NEIST, Jorhat – 785006 Assam

bksaikia[at]neist[dot]res[dot]in

There are other technologies developed at the lab and are on offer: https://www.neist.res.in/technologies.php

# Organic Superabsorbent 100% biodegradable polymer-Narayan Lal Gurjar/ EF Polymer, Udaipur

Narayan Lal Gurjar, a graduate from the College of Technology and Engineering, Udaipur, is the founder of EF Polymer Private Limited, a company specializing in creating sustainable agricultural solutions. Supported initially by NIAM with a grant of Rs. 25 lakhs and later by the OIST Innovation Accelerator with funding of 10 million Yen (approximately Rs. 65 lakhs), the company has been recognized for its innovative approach to water management in agriculture. Ankit Jain, a co-founder and graduate from Maharana Pratap University of Agriculture & Technology, Udaipur, also contributes significantly to the venture. The startup has been part of several accelerator programs and won the Social Alpha Agritech Innovations Challenge, supported by the Bill & Melinda Gates Foundation and Tata Trust in partnership with IIT Kanpur.

## Technology

Super-absorbent polymers (SAPs) are traditionally used in products such as diapers, sanitary towels, and cosmetics. These conventional SAPs are typically petroleum-based, derived from acrylic acid. In response to the growing demand for sustainable alternatives, SAPs made from natural raw materials like agricultural and shellfish waste, polysaccharides such as chitosan and cellulose, proteins, and poly(amino acids) have gained attention. A notable example is the creation of a biodegradable superabsorbent polymer from orange peels by Kiara Nirghin, which won the Google Science Fair Community Impact Award. This innovation involves cross-linking polymers through UV light and heat and using emulsion polymerization, resulting in a water-retaining hydrogel that acts as a reservoir in soil.

## Innovation

EF Polymer India has developed 'Fasal Amrit,' the world's first 100% organic bio-gel derived from fruit peels and biowaste collected from fruit juice shops. The innovation aims to enhance soil moisture retention and increase agricultural yields, especially in water-scarce regions. The biowaste is processed by drying in solar tunnel dryers and grinding into powder form. This powder is then used to extract pectin and cellulose, key ingredients for polymerization. When mixed with soil, Fasal Amrit significantly improves its water retention properties. This innovative product can absorb up to 50 times its weight in water and can repeatedly absorb and release water for six months. After its effective period, it decomposes and reintegrates into the soil within a year. The product's water-absorbing properties also prevent fertilizer runoff, making it more effective for prolonged use.

In addition to 'Fasal Amrit,' EF Polymer, in collaboration with Iwatani Corporation and Iwatani Materials Corporation, has developed Cy-cool™, a biodegradable ice pack made from natural materials like orange peels. After serving as a cooling agent, Cy-cool™ can be repurposed as a soil water retention aid, thereby offering dual functionality.

## Patents

Paper- https://www.researchgate.net/publication/344552472_Organic_Super_Absorbent_Polymers_SOP

ECO-FRIENDLY WATER RETENTION NATURAL POLYMER AND METHOD THEREOF, Publication number: 20220250033, Inventor: Narayan Lal GURJAR

## Commercialization

The company has offices in India and Japan.

https://efpolymer.com/technology

# Grey Water-Dr. Sonali Mokashi,/Cerulean Enviro Tech Pvt Ltd, Pune

Dr. Sonali Mokashi, a microbiologist with a PhD from Savitribai Phule University, is the Director of Cerulean Enviro Tech Pvt Ltd. Her entrepreneurial journey in sustainable waste management began with a water analysis laboratory and consultancy. The foundation for her current venture was laid during her project at MACS Agharkar Research on "Bioremediation of arsenic-contaminated groundwater." Building on this experience, she established a water testing laboratory, "Aquatech Solutions," where she conducted extensive experiments on greywater treatment and wastewater recycling. Her startup, Cerulean Enviro Tech, was incubated at the Center for Innovation and Business Incubation (CIBI) at the Indian Institute of Technology (IIT) Ropar.

## Technology

There are three primary treatment technologies for greywater treatment: physical, chemical, and biological. Besides traditional treatment methods, numerous advanced technologies have attracted interest in recent years, like membrane-based technology, improved electrocoagulation, nature-based solutions (constructed wetlands), and solar-based approaches. In India, the Central Pollution Control Board (CPCB) is the regulatory body dealing with laws and regulations for GW reuse, boron limitations of 2mg/L, sodium adsorption limits of 22mg/L, and electrical conductivity limits of 2250 µS/cm

## Innovation

The innovative technology is automated, compact, energy efficient waste water treatment plant with batch mode of operation giving consistent treated water quality output. The size is customizable varying from 1000 L per day to 500000 L per day capacity. Based on chemical treatment process with modifiable pre and post treatments. Multiple recycling (6 -7 times) of same water, compact size - 60% less space required than competitor, automated liquid chemical dosing, no dedicated operator needed, consistent treated water quality, 66% saving on fresh water purchase.

Vasudeva Aqa Greywater System collects the used water from the washing machine, kitchen sink, and bathroom except for toilet flush water. Thereafter, the water goes through a 4-step treatment process. Step 1 – this process removes the suspended impurities and heavy particles. Step 2 – this process separates the oil and grease. Step 3 - All the fine particle up to 3 to 4 microns are removed in this step. Step 4 – It used Carbon High-Density Filter to remove bacteria and odor.

NEERI offers Phytorid based system- https://kh.aquaenergyexpo. com/wp-content/uploads/2022/11/Grey-water-Treatment-and-Technologies.pdf

Guidebook on reuse of grey water in rural schools- https://sswm. info/sites/default/files/reference_attachments/NEERI%202007%20 Greywater%20Reuse%20in%20Rural%20Schools.pdf

## Patents

Indian Patent for the invention," A process for recycling domestic wastewater" 261355.

## Commercialization

Design and Manufacturing of ETP and STP, Design and Manufacturing of water softeners, DM Plants, UV plants, SCF Plants. Assembling UF, RO, ZLD plants. Operation and Maintenance services.

https://cerulean-envirotech.com/grey-water-treatment-plant/

# Iron Ore Extraction from Grinding Waste- A.LEO BERNARD, Ashok Leyland, Ennore

The team Prasanna Kumar, Plant Head and Leo Bernard manger in the corporate publication.

## Technology

The metalworking industry produces a large amount of waste material by machining. In contrast to conventional machining, the waste material generated during a grinding process is not purely metallic, but consists of abrasives, water, lubricants and metal chips. This mixture is called grinding sludge. Grinding waste is generated at soft stage, hard stage and heat treatment operations in truck production operations. Several recycling strategies in recent years, dealing with metallic waste products from different machining processes, have been proposed.

Paper from NML- https://docslib.org/doc/10891814/unit-processes-in-pyrometallurgy-drying-calcination-roasting-pelletizing-and-sintering

## Innovation

The firm wanted to convert waste into wealth, started investigating composition of grinding waste. The team selected Pyrometallurgical process and experimented with various stages with in-house team and with IITM. The process includes Conflagrating, Incineration, Smelting, Drossing, Sintering, Melting at high temp, Initially sludge is collected, dewatered and after demoisturization, decarbonation deoxidation the slag prepared with sintering and iron extracted.

https://ciisrehsawards.com/ehsr/ehsuploadpresdata/430_Environment_Restoration_-_Ashokleyland_Ennore_-_6273_0.pdf

The team also developed process to use grinding waste as replacement for sand in construction.

## Patents

Construction material comprising metaling grinding process waste as sand replacement, Indian patent 201841018186

## Commercialization

Processes for converting metaling grinding process waste as replacement for sand in construction blocks and for iron recovery are implemented in Ashok Leyland foundry and are available for transfer. For details and technology transfer contact:

A.LEO BERNARD, manager Plant engineering, Leobernard.a@ashokleyland.com

# Mixed Waste Conversion-Amardeep Singh/ Drossworld, Mohali

Amardep singh, graduate from Aligarh Muslim University is innovator and founder of Dross management Systems & Energy Solutions Pvt. Ltd.

## Technology

The government advocates segregation of waste as 1<sup>st</sup> step.

https://mohua.gov.in/upload/uploadfiles/files/Part2.pdf

## Innovation

Amardeep designed, developed and manufactured "DROSS-MAGIC" an instant un-sorted mixed solid waste disposal machine. It is really a new innovation and scientific approach which is the matchless in the field of Solid waste management and enables to adopting zero landfill mission. Dross MSW disposal technology is fully automated technological process that is able to process and covert MSW into high energy fuel pellets. The technology involves conversion of solid waste by an integrated process in a chronological and synchronized process of converting waste into useful products.

## Patents

SMART MACHINE FOR PROCESSING MULTIPLE TYPES OF GARBAGE TO YIELD COMMERCIALLY USEFUL PRODUCTS, WIPO Patent Application WO/2021/059292, Inventor- SHARMA AMARDEEP

The present invention discloses a smart machine for processing of unsorted garbage in a fast, efficient and economical manner. The machine uses almost 95% less water than conventional machines, which process garbage and use water for washing and cooling of the garbage processing machines. Further, the machine is highly energy efficient because it does not use any energy for heating and drying of the garbage. Rather the machine uses a novel and innovative mechanism of 'in-situ' heating, in which the heat generated during grinding using four specially designed blades (3 A) which help in the drying of the garbage and pulverizing it to produce a dry, fluffy powder which can be used to produce commercially useful products e.g. fuel, filler material etc. The vapors given off during the 'in-situ' drying process are condensed and re- circulated to cool the machine and other purposes, thus drastically reducing water consumption compared to conventional machines.

The machine consists of the following parts: a. Garbage receiving vessel: This is a large container of suitable capacity, made of suitable material such as mild steel/stainless steel and has a top lid b. Pulveriser: This is a mechanical assembly fitted inside the garbage receiving vessel and consists of specially designed four metal blades made up HARDOX 500 and 400 and D2 material. A metallic nanoparticle is sprayed on these blades which give more cutting efficiency and durability to these blades. All the four blades are fitted at an angle which is in the range of 30 DEGREE TO 60 DEGREE to each other. Pulveriser is operated with the help of electrical motor. Due to special blades and variable speed of rotation, a temperature in the range of 75 to 200 degree centigrade is achieved in garbage receiving vessel. The heat generated is not with the help of any external heating mechanism but only due to heat generated due to friction of specially designed blades, and the soring temperature helps in killing germs and pathogens, set at particular angle and rotated at particular speed under vacuum. Two operations are simultaneously carried out due to this single operation- One is pulverising of waste and other is heat generation. Further it is connected to a wireless modem for remote monitoring and control.

## Commercialization

https://drossworld.com/about-us/

# Refurbished Battery from Scrap-Rajat Verma/Lohum Cleantech, Noida

Rajat Verma, the founder and CEO of Lohum Cleantech Pvt. Ltd., holds a distinguished academic background with a degree from IIT Kanpur, an MS from Stanford, and an MBA from Harvard. Lohum Cleantech, based in Noida, specializes in recycling and refurbishing lithium-ion batteries, aiming to create a sustainable and circular battery materials supply chain. Under Verma's leadership, Lohum has been recognized by the Confederation of Indian Industry (CII) with awards for innovation in the SME manufacturing sector. Gazanfar Safvi, head of Recycling & Chemicals at Lohum, brings expertise in metallurgy and battery recycling.

## Technology

Lithium-ion batteries are widely used in various applications due to their high energy density and long cycle life. However, their longevity is affected by several factors, including battery chemistry, temperature, charging cycles, manufacturing quality, and more. After their primary use, lithium-ion batteries still contain valuable materials that can be recycled or repurposed. Efficient recycling processes and second-life applications for used batteries can significantly reduce the environmental impact of these batteries.

## Innovation

Lohum Cleantech has developed an innovative process to recover critical lithium-ion battery materials using its proprietary NEETM™ recycling technology. This multi-stage hydro-metallurgical technology involves using liquid solutions, such as acids, to extract refined battery metal salts with high recovery rates while minimizing CO2 emissions. The NEETM™ process ensures the high-yield recycling of lithium-ion batteries and contributes to an environmentally sustainable supply of raw materials. In addition, Lohum's approach combines second-life applications with recycling to enhance sustainability and value creation over the long term. The company has also pioneered reverse logistics in battery recycling, including the development of the DETX™ battery price index. This index provides transparency in scrap pricing by reflecting the live movement of material prices based on global demand and supply, helping stakeholders in the energy transition ecosystem manage price fluctuations. Lohum's process involves state-of-health testing and remaining useful life analysis to determine whether battery cells can be reused or need to be recycled. Lohum recycles all types of lithium-ion batteries, regardless of their cell chemistry and form..

## Patents

System for extracting electrode material from batteries, US20230076830A1, Inventor-Rajat VermaSyed Gazanfar Abbas SAFVIVikrant Singh

The present disclosure relates to a system for extracting electrode material from batteries. A shredding unit configured to receive the cooled feedstock from the freezing unit. The shredding unit is configured to shred the feedstock into powder form. A cyclone separator configured with the shredding unit, and configured to receive air bone electrode material particles generated as a result of shredding the batteries. A separating unit configured with the shredding unit, and configured to separate the electrode material particles. A cleaning unit operatively configured with the separating unit and the cyclone separator. The cleaning unit is configured to receive the powdered electrode particles from the shredding unit, and powdered electrode materials from a first output of the cyclone separator. A mixing agitator is configured to receive the powdered electrode material from the cleaning unit.

## Commercialization

https://lohum.com/scrap-battery-price-calculator/

# Rural Water Supply-Vinod Gaan/Rite Water Solutions, Nagpur

Vinod Gaan is the Chairman of Rite Water Solutions, a company dedicated to improving rural water supply systems in India. With over 40 years of experience in entrepreneurship and business management, Gaan has a background in chemical engineering from the Lakshminarayan Institute of Technology and has worked with major organizations such as Candy Filters India Ltd and the Steel Authority of India Ltd.

His son, Abhijeet Gaan, serves as the Director and CEO of the company and has a strong foundation in computer science engineering from BITS Pilani and an MBA in marketing from SP Jain, Mumbai. The firm has secured significant funding of Rs. 100 crores from the Water Access Acceleration Fund (W2AF) to expand its clean water initiatives

## Technology

Rite Water Solutions employs innovative technology such as Water ATMs. to address the scarcity of clean drinking water in rural areas. Water ATMs. are automated water dispensing units that operate 24/7, often powered by solar energy. These machines can purify water through solar-powered osmosis or be connected to the grid system. The water ATMs. are designed to be accessible and user-friendly, with water cards that users can top up with credit to obtain water at affordable rates. In Uganda and Bangladesh, for example, water ATMs. dispense water for a fraction of a cent per liter.

## Innovation

Rite Water Solutions has been at the forefront of initiatives like the Jal Jeevan Mission's "Har Ghar Jal" campaign and has collaborated with international partners like Johkasou Technology from Japan. The company's Water ATMs. and mobile water solutions have successfully provided clean drinking water to over 1.5 million people across India. Their technology effectively removes harmful contaminants from water, improving the quality of life in both rural and urban communities.

The Water ATM system functions as a decentralized community water purification plant based on Ultrafiltration, Reverse Osmosis, and Ultraviolet technology. It uses existing water sources such as borewells or overhead tanks to provide high-quality drinking water at a minimal cost of 25 paise per liter, compared to the much higher prices of bottled water. Each household is provided with an RFID-based Water ATM card and a 20-liter can for collecting water. The system validates the user's identity and card balance, dispensing the required quantity of water automatically. This system ensures continuous access to safe drinking water around the clock, significantly reducing the use of plastic bottles and providing employment opportunities for local youth..

## Patents

Coin operated water vending machines, GB1212382A

## Commercialization

Rite Water Solutions (I) Pvt. Ltd. is an ISO 9001:2008 certified organization specializing in providing solutions for Rural and Urban Water Quality Improvements.

Specialties: Electrolytic Defluoridation Plant, Reverse Osmosis Plants, Nano Media Technology for Fluoride & Arsenic Removal, Electro Chlorination Plants, Online Water Quality Monitoring Systems, Nitrate & Iron Removal Systems, and Water Softening Plants.

# Silica Plastic Block-Manish Kothari, Rhino Machines Pvt Ltd, Anand

Manish Kothari, the Managing Director of Rhino Machines Pvt Ltd, has spearheaded significant advancements in sustainable construction materials. With a background in mechanical engineering from Maharaja Sivaji Rao University, he joined the family business, Rhino Machines, which was founded by his father, R.C. Mothari, an experienced foundryman and alumnus of IIT Bombay. Rhino Machines began as a project consultancy and evolved into a manufacturer of cutting-edge foundry equipment, earning accolades like the Innovation Award finalist spot at the Global Cleantech Innovation Programme supported by UNIDO-MSME-FICCI-Cleantech Open-GEF in 2015.

## Technology

Rhino Machines has developed an innovative solution to repurpose scrap plastic waste (SPW) and foundry sand (FS) into sustainable building materials, specifically bricks. The process involves creating bricks with varying ratios of FS and SPW (20%, 30%, 40%), which are then tested for durability and strength. Studies have shown that these bricks exhibit 85% greater strength than traditional fired clay bricks. A ratio of 70% FS to 30% SPW provided the highest strength values. The SPW bricks demonstrated superior compressive strength (29.45 MPa) compared to fired clay bricks (14 MPa) and showed excellent resistance to water absorption and acid exposure, making them highly durable and suitable for various environmental conditions.

**Innovations:** The firm identified a problem and successfully addressed $CO_2$ gas waste produced in the sodium silicate bonded sand process in the foundry industry. The moulding process was cheaper and used extensively in India, but lacked solutions to reclaim sand after its first use. They developed machinery and process for recovery of moulding sand, thermal sand and core stand. They tied up with Fata, Italy for Green sand (water bound bentonite/clay) reclamation process and during implantation they came across another issue, residual dust. While system could recover 70-80% of the material, 20-30% waste dust was left behind which had disposal challenges. Being less than 100 micron in size, the dust could fly when dumped and add to the pollution problem. R&D team experimented and established a bonding of unsegregated plastic waste. This constitutes polyethylene, polypropylene, hdpe, ldpe and a difficult multi-layer plastic package in proportion of about 20% to 30% (tested up to 50%) with dust waste from foundry and the discarded sand waste. The SPB (Silica Plastic Block) is produced from 100% waste and can be recycled and moulded again.

Licensed technology to start-up Upcycle Chakra for deployment and implementation, Capacity from 30 kg/hr to 4 t/hr.

## Patents

CO2 Sand Reclamation Machine.CO2 Sand Reclamation Machine, IN 203200 · Issued Nov 7, 2002

## Commercialization

The company designs and supplies several plants; Multiflex - High Pressure Moulding Machine, Ecoflex - Sand Handling System, RTM - Rotomax Mixer: RMC - RTM Mixer Cooler, RTC - Online Sand Controller, S1 - Mechanical Scrubber, S2 - Thermal Sand Reclamation, S3 - Green to Core Sand Reclamation.

https://sites.google.com/view/rhino-machines-pvt-ltd/home

# Smart Self-Healing Coatings-S K Dhawan/ Vikas Group, NewDelhi

Dr. S.K. Dhawan, a distinguished scientist with an extensive background in chemistry, is currently the Managing Director of Vikas Lifecare Limited, New Delhi. With a career that spans roles as an Emeritus Scientist and Chief Scientist, Dr. Dhawan has focused on areas such as conducting polymers, smart coatings, waste management, and nano composites. He has published 180 scientific papers and holds several patents, including 8 US and 8 Indian patents. His innovative work in self-healing coatings has garnered him multiple awards, including the DST-Lockheed Martin Innovation Award in 2014 and the ASDF Global Award in 2015.

## Technology

In past few decades lot of research has been carried out on conducting polymer-based coatings. Conducting polymers have been used for anticorrosion application either by electrodepositing it on metal surface or by direct addition to the corrosive medium as an inhibitor. Various conducting polymers like Polypyrole, polyaniline, and polythiophene etc have been studied for anticorrosive properties. Among the various conducting polymers polyaniline and its derivative exhibit excellent anticorrosive properties. This conducting polymer shows dual protection mechanism by providing anodic protection to underlying metal and shifting its potential to passive region. Secondly it acts as a barrier against highly corrosive conditions.

## Innovation

The present innovation deals with the designing of smart self-healing coatings of conducting polymers which can be used for preventing corrosion of iron under hostile environmental conditions. The aim of the invention is to design conducting polymer composites by incorporating filler materials and suitably selecting a medium for polymerization so that the resultant epoxy coatings can be used for prevention of corrosion in saline water conditions. Aim was to design conjugated polymers which show Smart action and have self-healing ability. Self-healing mechanism arise due to the redox property of the conducting polymer which leads to pin-hole and scratch free passivation. These coatings are environmentally friendly/ based on green technology (free from heavy metal ions and hazardous chromates), have long service life & economic feasible. These polymers can also be used as additive formulation with cement for the water-soluble conducting polymer based antirust cement. Main aim of the study is to design conjugated copolymer in a suitable polymerization medium and blended with epoxies so that these conjugated copolymer blends when coated on a mild steel surface, can be used for the protection of iron and mild steel under corrosive conditions.

We designed conducting polymers and incorporate a suitable composition in the paint matrix which helps in self-healing mechanism of paint surfaces coated on metals if a **scratch appears on the surface.**

## Patents

Technology has been developed by us on "Designing of Smart Intelligent Nanocomposites for corrosion protection of iron & mild steel in saline environment". A patent has been granted by Indian Patent Office, Govt. of India, Patent No. 480897 on December 12, 2023.

## Commercialization

Vikas Lifecare Limited is a leading provider of high-end specialty chemicals to customers across the world. We are passionate about finding innovative solutions to the most pressing environment and safety challenges of our times. https://www.vikaslifecarelimited.com/rnd/

# Waterless Urinal-Neha Bagoria/Tapu Sustainable Solutions, Mumbai

Neha Bagoria, founder of Tapu Sustainable Solutions, is a pioneer in sustainable sanitation technology. With a Master's degree in Computer Science from Rohtak University and a PG Diploma from National Law University, Neha was supported by the Department of Scientific and Industrial Research (DSIR) under the PRISM scheme to develop her innovative retrofit waterless urinal kit, 'EcoTrapIn.' Her work has been recognized at International Innovation Fairs organized by the Indian Innovators Association in Bengaluru, Hyderabad, and at the Korean International Women's Innovation Exposition.

## Technology

Waterless urinals operate by diverting urine through a one-way valve into a trap that contains a chemical sealant. This sealant, which is less dense than urine, creates a barrier that prevents odors from escaping back into the restroom. As urine passes through the trap, it displaces the sealant and is then directed to a drain connected to the sewer system. Over time, the sealant must be replaced to maintain performance and prevent odors. The maintenance interval typically depends on the usage frequency, generally around every 1,500 uses.

The key component of waterless urinals is the trap, a removable cartridge that facilitates waterless drainage. The trap is designed with a central reservoir and a series of slits that allow urine to flow into the cartridge and through the sealant. Once the reservoir reaches a certain level, the excess liquid drains out through a central pipe connected to the waste line.

## Innovations

EcoTrapIn Waterless Urinal: EcoTrapIn® (patent technology) is our 3rd version waterless urinal envisioned to save 1,67,900 liters of potable water per urinal annually with substantial cut in utility bills and dry hygienic restrooms. The product technology retrofit design converts conventional urinal into waterless urinal. EcoTrapIn® waterless urinal is a cost effective (no payback time) solution designed with simple 2 step installation, maintenance free, no clogging and no bad odour to enhance hygiene in male urinals. Utility bills are slashed by 90% and water is saved by 95% with the usage of product. EcoTrapIn® waterless urinal is proposing the most cost effective, ergonomic and green product that converts conventional urinal into dry hygienic waterless urinal with low shifting cost (no replacement) of urinal bowl like most other waterless urinals.. Further, two versions EcoTrapInPlus® and EcoTrapInXtra® has been developed and launched into market. No-flushing DryTrap technology prevents reaction between urine and water, hence urinal is free from foul smell, scale formation and drain pipe blockages. The touch free green product cost Rs. 18/day with the savings of 90% in utility bills viz.,water charges, cleaning agents, plumbing/repairing service,electricity/battery etc.

## Patents

A Urinal System, application number-201621007814, Inventor- Neha Bagoria

A URINAL SYSTEM A urinal system includes a cartridge to be fixed in a channel of urinal flow as a part of a urinal, and to allow flow of the urinal, a flap placed inside the cartridge such that to be either in a close position to close the flow path of urine or to be in an open position to open flow path of the urine, and an actuating mechanism to effectuate the flap to be either in the close position or the open position.

## Commercialization

http://tapu.co.in/#

# Innovations profiled in 2022

## GROUP C- WATER & SANITATION

1. Low-Cost Arsenic Removal Filter for Drinking Water- IIT Kharagpur

2. Ground Water Detection- WaterQuest

3. Atmospheric Water Generator (AWG) with Remineralization- Maitriaqua

4. Sewer Cleaning Robot-Gen Robotics

5. Fertilizers from Urine- Waste Chakra

6. Waste Water Treatment- Indrawater

7. Water management- Kritsnam Technologies

8. Water testing-Elico

9. E.coli Test Kit- Earthface Annalytics

10. Recirculating Aquaculture Systems (RAS)- Oriental Aquamarine Biotech

11. Pipeline Inspection robots-SOLINAS

12. Textile Reinforced Concrete Prototyping Technology -CSIR -SERC

13. Colourful Pavement Tiles from Waste Plastic-CSIR-NPL

14. Extraction of Cobalt metal/salt from the black powder of Li-Cobalt batteries- CSIR-NML

15. Oneer Water Disinfection System-CSIR-IITR

# Innovations profiled in 2023

## GROUP D: WATER & SANITATION

1. Artistic Air purification Tower - Verto-Studio Symbiosis, New Delhi

2. Electrolytic De-fluoridation (EDF) Technique- NEERI Nagpur

3. e-Waste recycling- Ecoreco, Mumbai

4. HeliborneTEM technology for aquifer mapping- CSIR-NGRI, Hyderabad

5. Hot water dissolvable sanitary napkins, Cresa Greentech, Pune

6. Hydrogen Electrolyzer- Newtrace Pvt. Ltd.

7. Hydrogen Fuel Cell bus- Sentient labs/ KIIT, Pune,

8. Intelligently Stirred Thermophilic Anaerobic Reactor- Sankar Ganesh Palani, BITS Pilani, Hyderabad campus

9. Marine Oil Spill Remediation- NIOT, Chennai

10. Moving Bed Biofilm Reactor (MBBR)- NIT Warangal

11. Muffler and Particulate Separator for Internal Combustion Engine Exhaust- RnA Vortech Innovations, Chandigarh

12. Plastic Waste into Green Fuel and Energy- DBT-ICT Centre for Energy Biosciences

13. Sanitary pad waste recycling- Pad Care Labs, Pune

14. Water Management Platform (MIDAS)- Ekatvam Innovations, Thane

15. Waterless urinal- Ekam Eco Solutions, New Delhi

# Industrial Solutions (2024)

# 3D Printed Concrete Bridge-Amit Ghule/ Simpliforge Creations/IITH

Amit Ghule, a graduate of MCT's Rajiv Gandhi Institute of Technology with a Master's from Northeastern University, is the founder of Simpliforge Creations, a startup specializing in 3D concrete printing solutions. Waseem Chaudhary, a graduate from M.H. Saboo Siddik College of Engineering, serves as the Chief Technology Officer (CTO). The company collaborates closely with Prof. Kolluru V.L. Subramaniam, a professor in the Department of Civil Engineering at IIT Hyderabad (IITH). Prof. Subramaniam is currently engaged in a project titled "Development of Integrated 3-D Reinforced Sandwich Structural Panels for Rapid Construction of Building Systems," funded by the Uchchatar Avishkar Yojana, Ministry of Human Resource Development.

## Technology

Construction 3D printing involves creating construction elements and entire structures layer-by-layer using a 3D printer with materials such as concrete, geopolymers, and other construction materials. This technology provides architects and engineers with the freedom to design unique and innovative structures while offering advantages such as cost-effectiveness, agile supply chains, optimal resource utilization, and decentralized manufacturing. The concept of 3D printed concrete originated at Rensselaer Polytechnic Institute (RPI) in New York, where Joseph Pegna first applied additive manufacturing techniques to concrete in 1997.

## Innovations

Simpliforge Creations has successfully deployed a 3D printed pedestrian bridge on the IIT Hyderabad campus. The bridge's concept and design were developed and evaluated by Prof. K.V.L. Subramaniam and his research team from the Department of Civil Engineering at IIT Hyderabad. The bridge, spanning 7.50 meters, was printed off-site by Simpliforge Creations and later installed after a smaller prototype was load tested. The design of the bridge focuses on form optimization to minimize the use of concrete and reinforcement, following the principle of "Material follows Force." This principle, along with stress analysis, guided the determination of the reinforcement and shape of the bridge. The material developed by Simpliforge Creations was rigorously tested for the required rheological performance. The project demonstrates several advances in material processing and design methodology, showcasing the potential of 3D concrete printing technology for rapid and efficient construction.

## Patents

V.V. Rangarao, K.V.L. Subramaniam, S. Suriya Prakash 'Lateral Reinforcement System and Method for Concrete Structures', Patent Number: 3001/ CHE/2015 dated 16 June 2015

V.V. Rangarao, K.V.L. Subramaniam, S. Suriya Prakash 'Lateral Reinforcement System and Method for Concrete Structures', US Patent No. 9719245 B1, Aug. 1, 2017

## Commercialization

Simpliforge's Construction 3D printer is recognized as India's first and South Asia's largest robotic construction 3D printer. The printer is capable of producing a variety of elements, including landscaping features, furniture, statues, wall facades, and full-scale houses. Simpliforge has developed its own 3D printable construction material, which reduces the use of traditional cement-based materials. By combining proprietary software and materials, Simpliforge positions itself as a significant player in the global 3D printed construction domain. The company's 3D printing systems can construct structures with a range of 8.5 meters at printing speeds of up to 500 mm/s.

https://www.simpliforge.com/

# Antimicrobial Solutions-Anasuya Roy/ Nanosafe solutions/IITD

Dr. Anasuya Roy is the founder and CEO of Nanosafe Solutions, a technological startup incubated at IIT Delhi. She completed her PhD at the Indian Institute of Technology Delhi in 2019. Dr. Roy is a recipient of the Biotechnology Ignition Grant (BIG) from BIRAC, DBT in 2019 and was awarded a DAAD fellowship for the IIT-DAAD Master Scholarship program. She has contributed 11 published papers in peer-reviewed journals and authored three book chapters. Nanosafe Solutions, under her leadership, focuses on developing antimicrobial functionalized polymer nanocomposite materials as base materials for manufacturing containers and bottles. Professor Mangala Joshi, co-founder and Technical Director of the startup, is a renowned expert in polymer science and has spearheaded multiple projects, including a BIRAC-funded initiative to develop antimicrobial water storage containers.

## Technology

Copper nanoparticles have been explored for their antimicrobial properties, particularly in enhancing the quality of protective clothing and other materials. Research has demonstrated that copper nanoparticles, when embedded in polymer matrices like polyvinyl-methyl-ketone, PVC (polyvinyl chloride), and polyvinylidene fluoride, exhibit strong biostatic activity against microbes such as Saccharomyces cerevisiae, Escherichia coli, Staphylococcus aureus, and Listeria species. Nanosafe Solutions has leveraged this technology by integrating copper nanoparticles into polymer nanocomposites to create antimicrobial products.

## Innovation

Mixing traditional science with nanotechnology, IIT Delhi-incubated startup Nanosafe Solutions has developed a range of antimicrobial i.e. antiviral, antibacterial and antifungal water storage containers and launched it as "AqCure", which is based on the inherent antimicrobial properties of copper. AqCure is a patented technology in which active nano-copper is released in a sustainable manner from a polymer matrix. The released copper makes the outer and inner surface of the container antimicrobial, reducing transmission of microbes upon direct contact, and making the stored water microbiologically safe. Additionally, the released copper in water is within permissible limits and thus fortifies stored water as copper is also an essential micronutrient, needed for growth.

Elaborating the features of AqCure, Dr. Anasuya Roy, an IIT Delhi Alumna and CEO, NanoSafe Solutions said, "AqCure water containers have>99.99% antibacterial, >99.99% antifungal, and >99% antiviral activity tested as per ISO and ASTM standards. These containers are made from BPA/BPS-free high quality food grade polymers and is ideal for home and office use."

## Patents

Antimicrobial Non-Woven Fabric for Safe Water Filtration, IN 201911018771

## Commercialization

The IIT Delhi startup "Nanosafe Solutions" has launched an antimicrobial and washable face mask "NSafe", which is reusable up to 50 launderings, thus greatly cutting down the cost of use. New product, AqCure water storage containers are available in different size variants from 700 ml office bottles to 1-liter refrigerator bottles to be used at homes to 10-20-liter bubble tops and cans used for storage and distribution of potable water. AqCure polymer masterbatch (polymers compounded with active nano-copper) granules based on different carrier polymers are also available, which can be used in polymer molding and extrusion operations to make the final products antimicrobial.

https://nanosafesolutions.com/

# Cloud Communication-Dasari Uday Kumar Reddy/Tanla platforms, Hyderabad

Uday Reddy MBA from University of Manchester is CMD at Tanla Platforms ltd. A first generation entrepreneur in the space of telecommunication, Managing director and Chief Architect of the company, driving the meteoric rise from a products-based solution provider to one of the largest publicly traded Business Cloud Communications Company. Tanla is now recognized as the largest and fastest-growing business cloud communications provider and was most recently ranked as one of the top "1000 High-Growth Companies in the Asia Pacific" by Financial Times (FT). Gartner recognizes Tanla in the latest CPaaS market guide as a key global solution provider in the Developer market; the only Asian Company to be recognized under the Developer category.

## Technology

The need for contextual communications is one of the largest emerging trends in customer-facing communications. Cloud-based solutions such as CPaaS (Communications Platform as A Service) are leading the charge towards next-generation communications by allowing enterprises of any size to easily develop and embed communications features without needing to build backend infrastructure and interfaces. Businesses are turning to Communications Platform as a Service (CPaaS) more than ever to improve and streamline their customer experience. This technology that simplifies communication integration from voice, messaging and video is transforming the way we interact with customers in an increasingly digital age. Without strong security controls, CPaaS companies are vulnerable to fraud attacks. In some cases, millions of SMS text messages have been leaked online. These data breaches have led to loss of trust and revenue for even the largest organizations.

## Innovation

Unwavering commitment to minimizing spam, fraud, privacy breaches and any other kinds of Unsolicited Commercial Communication (UCC) led to development of Trubloq – the world's first block chain enabled communications stack that combats UCC. It is the single-largest deployment of block chain powered use case processing to the tune of 350 billion transactions per year. A whopping 70% of India's A2P SMS traffic is processed by Trubloq and it hit the incredible milestone of scrubbing 1.1 billion messages in a single day in 2021. Rich Communication Services (RCS) is a protocol created to replace SMS messages with a text message system with several app-like features – read receipts, engagement insights and sharing multimedia, locations etc.

## Patents

AUTOMATED METER READING SYSTEM AND METHOD THEREOF, Publication number: 20090146839, Inventors: Dasari Uday Kumar Reddy, Kathirisetti Satish, Chachan Navnit

## Commercialization

Tanla was founded in 1999 as a small mobile messaging company by Uday Reddy with a bunch of mobile messaging experts who set out to create a world-class messaging service. In 2021 partnered with Microsoft to launch block chain-enabled CPaaS platform Wisely, Partnered with Truecaller to offer Business Caller ID solutions to enterprises.

https://www.tanla.com/platform/wisely

# Digital management system-Siva Rama Brahmam/Jaajitech.com, Ongole

Kondapi V. Siva Rama Brahmam, a Chemical Engineering graduate from Andhra University, is the CEO of Jaajitech, a company focused on digital transformation technologies for the process industry. He has a rich background in working with Indian Oil Corporation (IOC) and the ATME Group in Kuwait, specializing in Advanced Process Control (APC) projects. Dr. Phaneendra Kondapi, who leads the USA operations of Jaajitech, has an extensive background in subsea engineering, production management systems, digital technologies, and operator training simulators. He holds a Ph.D. from Tennessee Technological University and has also completed his B.Tech. and M.Tech. at Andhra University. Jaajitech has been recognized multiple times for its innovative solutions, including being shortlisted by CIO Tech Outlook as one of the '10 Most Promising Energy and Utility Tech Companies' in 2022 and 'Company of the Year' in 2021.

## Technology

Jaajitech focuses on the integration of data historian technology and digital twins to optimize operations in industries such as oil and gas refineries. Data historians are time-series database applications developed in the late 1980s to collect and store data related to industrial operations. These technologies have evolved to support the digital transformation of refineries, allowing for the digitization of processes, improved production, reduced operational costs, minimized risks, and enhanced decision-making capabilities related to asset maintenance and operation.

## Innovation

Jaajitech's flagship product, inSis Soft Gateway, is a software solution that connects with existing PLC/DCS/SCADA systems to read real-time operations data. This data is securely stored in the inSis Luna Cloud. The inSis Manufacturing Information System and Operational Excellence provide real-time data visibility with role-based dashboards, allowing for monitoring, data analytics, and troubleshooting of operational issues. Automated web-based Excel reports provide timely and accurate data accessible via mobile devices. Additional solutions include the Remote Oil Well Monitoring Solution, which enables remote monitoring of operations data from oil fields and related assets, and the Emission Management Solution (EMIMS), which facilitates the collection and collation of emission data from online analyzers.

## Patents

IOT and patents- WIPO database:

> https://www.wipo.int/edocs/plrdocs/en/internet_of_things.pdf

## Commercialization

Jaajitech, provides an integrated platform for the digital transformation of process industry. inSis Suite, an award winning software, features four pillars of digital transformation for manufacturing execution,

1. Operations - DOES: Digital Operations Execution System & Connected-worker platform (Digital Operator Logbooks, IOW & Safe Limits, Morning Meeting Dashboard, Safety Interlock Bypass Management, AMADAS - Analyzer Maintenance and Data Acquisition System etc.)

2. Process - PMS: Performance Management System (Data Historian, Analytics, Dashboards, KPIs, Reports & Mobile Apps)

3. Maintenance - AP360: CMMS, Digital Checklists & Asset Performance

4. Quality - QM: LIMS, ELN & Digital Laboratory

> https://www.jaajitech.com/Company/AboutUs

# Electrolyser-Siddharth Mayur/ Homihydrogen, Pune

Siddharth Mayur, a graduate from Sydenham College of Commerce & Economics, is the Founder and Managing Director of h2e Power Systems Private Limited and the Founder & Executive Director of homiHydrogen Private Limited. homiHydrogen Pvt Ltd is a joint venture promoted by h2e Power Systems Private Limited and BlueBasic AMA Engineering Private Limited, an Indo-Italian company focusing on sustainable mobility and green technologies. The company is involved in three verticals in India: automotive emission solutions, cryogenic tanks for long-range transportation, and the manufacture of electrolyser technologies. Dr. Prashant Karkal Kamath, a managing partner of DentaCare, is also associated with this venture. Greenstat Hydrogen India Pvt Ltd, a part of the Norwegian Green Energy and Technology company, The Greenstat Group, collaborates with homiHydrogen to advance green hydrogen initiatives in India and overseas.

## Technology

Green hydrogen is typically produced by water electrolysis. In this process, water is split into hydrogen and oxygen by an electric current. Different technologies exist to produce hydrogen by electrolysis, including the four best recognized: Anion Exchange Membrane (AEM), Solid Oxide Electrolysis Cell (SOEC), Polymer Electrolyte Membrane (PEM) and Alkaline Water Electrolysis (AWE). Alkaline Membrane Solid Electrolyser technology is the most innovative and still relatively unknown. The basic principle and process scheme are the same as PEM technology, but in this case, charge transport is guaranteed by OH- ions as in traditional alkaline technology, combining the advantages of the latter (mainly the presence of more affordable components) with the advantages of PEM technology (mainly the possibility of obtaining high-pressure hydrogen).

## Innovation

H2 Energy Italy offers electrolyzers based on PEM, alkaline and our proprietary AMSE® technology, manufactured in Italy with Italian technology. These electrolyzers ensure a 99.999% purity of H2 and can achieve up to 80% efficiency under optimal conditions. This results in a hydrogen production rate of up to 200 Nm3/h*MWe. AMSE Electrolysers combines features of PEM and AEM.

## Patents

APPARATUS FOR GENERATING HYDROGEN AND OXYGEN THROUGH ALKALINE ELECTROLYSIS, AND CORRESPONDING PROCESS, WO2021214318A1, Inventors-VALLI ALBERTO [IT]; CASTIONI EMANUELE [IT], Assigned to SPI Consulting.

## Commercialization

Under the Strategic Interventions PLI program of Government of India, in the category for using domestic technology stack, setting up Manufacturing Capacities for Electrolysers. HomiHydrogen Private Limited won bid for 101.5 MW capacity which will provide them maximum incentive of Rs. 150,22 crores,

**https://homihydrogen.com/technology.html#TechnologyComparison**

# Gujfly Magnetic Flywheel-Liyakatali Babi, Mehsana, Gujarat

Babi Liyakat is Innovator at Mechanical Engineering Sciences Laboratory (MESL) and Managing Director of Babi Magnetic Flywheel Energy Pvt Ltd.

## Technology

The early Villiers engines were all equipped with horseshoe pattern flywheel magnetos, which at that time were principally produced in Germany. During the war period, however, the firm started to manufacture its own magnetos, and after careful consideration it was decided to adopt a flywheel type, which, as the engine was already fitted with a flywheel which would house it, saved a projecting bracket, a chain, and two driving sprockets. After extensive experiments, the Villiers flywheel-type magneto was adopted in 1919 as standard equipment on all engines, and is now supplied for all British outboard-type marine engines, and for several Continental makers of similar motors.

## Innovation

Babi Magnetic Flywheel Energy Pvt. Ltd. owner Babi brothers came up with a unique concept of magnetic flywheel. His innovation is basically a magnetic wheel attached to the flywheel. Through the rotating magnetic field of the attached wheel, the wheel of the vehicle rotates even without fuel for some distance.

## Patents

KINETIC MOTION ENERGY BY ROTATING MAGNETIC FIELD IN WHICH FLY WHEEL REDUCES LOAD AND GENERATES EXTRA POWER TO MOTION. Patent number 369155, Innovator- Ayub Khan M Babi

## Commercialization

Address: MEHSANA, GUJARAT, INDIA, PIN:384001, ❖ Mobile: +91 96242 76786,

❖ Email: infogujfly@gmail.com, ❖ Website: www.gujfly.com,

❖ Youtube: https://www.youtube.com/@gujflyindia/videos

# High Energy Material-Dr. GM Gupta/Global Engineers, Delhi

Dr. G..M.Gupta), a mechanical engineering graduate from Aligarh Muslim University, began his career with DCM where he worked for 20 years and after a brief stint in USA at Pittsburg Steel he returned to India to start Global Engineers Ltd. In recognition to his contribution, the company has bagged 2 National Awards, 5 State Awards for Quality, Productivity and Entrepreneurship in the Engineering sector and IEI industry excellence award 2015 in manufacturing and processing.

## Technology

Two types of solid propellants (homogeneous and heterogeneous) are distinguished by their constituent ingredients and the condition in which they are linked. In a homogeneous propellant, the ingredients are linked chemically and the resulting physical structure is homogeneous throughout. Typical examples of homogeneous propellants are single-base (NC and additives), double-base (NC, NG and additives), and triple-base (NC, NG, NQ, and additives) propellants. The main ingredient in a single-base propellant is NC, gelatinized with ethyl alcohol as solvent. NG is made by acid nitration of natural cellulose fibers from wood or cotton and is a mixture of several organic nitrates. Small amounts of chemical stabilizer and flame suppressant are also added. The propellant grain is coated with carbon black to keep the surface smooth. At High Energy Materials Research Laboratory (HEMRL) a number of single base compositions have been developed by solvent process for a variety of applications in small arms and gun ammunition. These consist of nitrocellulose with other ingredients like modifiers and stabilizers to control burn rate and other properties. They are extruded into tubular, multi-tubular or honeycomb shapes or spherical balls. These propellants are used in all Indian Small Arms Systems.

## Innovation

Single Base Propellant is high energetic material essentially required to produce ammunition for Bofor Guns etc. and imported from Russia. GOI selected the firm to setup manufacturing facility at Ordnance Factory Bhandara.

Nitrocellulose is another high energetic material used for manufacture of all kind of ammunition. The firm together with an Austrian company innovated the technology and set up facilities at Ordnance Factory Nalanda and Ordnance Factory Bhandara to produce NitroCellulose successfully. Turn key projects for manufacture of explosives and propellants produce highly toxic effluents. The company developed a unique Effluent Treatment Plant which not only removes toxic components but also recycles the treated effluent.

## Patents

Method of producing progressively burning artillery propellant powder and agent adapted thereto, Patent number: 4654093, Assignee: Aktiebolaget Bofors (Bofors), Inventors: Boo Bolinder (Karlskoga), Hermann Schmid (Karlskoga)

## Commercialization

Global Engineers has collaborations with world's leading companies for setting up turnkey projects & equipment for Explosives and Propellants Plants for Ordnance factories such as:

Nitrocellulose (NC) Plants: Initiating Explosives (Lead Azide, Lead Styphnate, Tetrazene, Sodium Azide, Styphnic Acid) Plants

Fine RDX Plants: NitroGlycerine, NitroGuinodine, and Nitroglycols, Dynamite, Propellants, Ammunition Disposal Plants, Single Base & Multi Base Propellant Plants

Ball Powder Plants, Bleached Linter Manufacturing Plants

Tear Gas Manufacturing plant

http://www.globalengineers.co.in/Explosives.html

# Hydrogen Powered VTOL-Maruthi Amardeep Sri Vatsavaya/Bluj Aerospace, Hyderabad

Maruthi Amardeep Sri Vatsavaya is Co-Founder/CEO. An entrepreneur with a deep passion for technology and organization building. University of Cincinnati alumnus with more than 15 years' experience building rockets & aircraft. A graduate from Institute of Aeronautical Engineering, Dundigal, starting as a Student intern at NAL, he garnered valuable experience at GE Global Research, Bengaluru, B/E Aerospace, U.K Collins Aerospace, Hyderabad before joining the startup Sky Root Aerospace, Hyderabad. Utham Kumar Dharmapuri is Co-Founder and CTO, also a graduate from Institute of Aeronautical Engineering, He did MTech from JNTU worked at Cyient and Boeing. Pardmashree awardee Gnanagandhi Vasudevan with 40+ years of experience in Rocket propulsion at ISRO, pioneer in Cryogenic technology who also has been instrumental in developing India's 1st Hydrogen electric bus is their Chief Consultant- Propulsion. **Technology**

Swiss aviation startup Sirius Aviation AG brings out Sirius Jet, the dubbed world's first hydrogen-powered VTOL aircraft that is co-designed by BMW's Designworks and Sauber Group. Hydrogen-electric program builds on technology developed by Joby subsidiary H2FLY, acquired in 2021, and forms part of Joby's future technology roadmap; Joby's hydrogen-electric, vertical take-off and landing demonstrator aircraft completes landmark 523 mile flight, with water as the only by-product;

## Innovation

The startup plans to build a 10-seater hydrogen-electric passenger aircraft that has a range of up to 1,000 km, by 2028.

## Patents

High efficiency hydrogen fueled high altitude thermodynamic fuel cell system and aircraft using same, US11565607B2, Assignee Joby Aviation.

## Commercialization

The startup secured $2.25 million in funding led by Endiya Partners. Another company in this space is ePlane, which has raised $6.47 million from investors such as Speciale Invest, Micelio, and 3one4 Capital. The ePlane Company is co-founded by IIT-M professor Satya Chakravarthy and his student Pranjal Mehta. They are currently testing a scaled-down prototype and expect to have their first cargo plane ready as early as next year.

https://blujaero.com/

# Industrial Design Consultancy-Prof. Kishore Munshi/Ctech lab/IITB

Prof. Kishore Munshi is Managing Director at CTech Labs. Strong arts and design professional with a Master's degree focused in Industrial and Product Design from IIT Bombay / Royal College of Art, London. Design consultant to many and author of popular book- Mind into Matter Reflections on Design Characteristics, Technology and Society https://books.google.com/books/about/Mind_into_Matter.html?id=LtQBEQAAQBAJ

He was UNESCO Fellow at Technical University of Hanover, Germany, de Technologie de Compiegne, France, and worked with some of the leading design groups in Europe - Sottsass Associati, Milano, Italy. ENFI Design, Paris, France, Lindinger & Partner, Hanover, Germany etc. Professor and former Head of the Department, Industrial Design Centre, IIT Bombay, Professor of Industrial Design at Oslo School of Architecture & Design, Norway, where he started the Masters program in Industrial Design. Professor of Design Management at Middle East Technical University, Turkey. Visiting Professor at Lulea University of Technology, Sweden. Management adviser and R&D (Product Development) Consultant to more than 50 industrial organization including L&T Ltd., IBP Ltd., Godrej & Boyce Ltd., Ponds India Ltd., HMT Watch Directorate, Crompton Greaves Ltd., Godfrey Philips Ltd., DRDO Labs, Indian Railways etc.

## Technology

Examples of designs from IDC, IITB- 'Zero Sum' - An educational game based on maths, designed by Industrial Design Center was launched by Gakken, Japan on 1st May in the presence of Indian Ambassador, Gakken officials and other dignitaries. Zero Sum was designed by students Priya Ganadas, Chandni Kabra and alumnus Ashish Ganu under the guidance of Prof. Athavankar.

## Innovation

To promote employment and development in rural areas, CTech offers to collaborate with NGOs & rural based organizations to develop special products for rural communities & industries. CTech has developed LED solar lighting systems for rural sector applications - Jugnoo LED Lights eLantern, eCandle, Hawker Light and low weight Cap Light for miners. CTech in collaboration with IITB has designed and developed sustainable and eco- friendly energy saving electrical vehicles – two wheelers, three wheelers for interior use - airports, shopping malls, IT campuses, railway stations, factories etc. CTech strategy is based on niche EV marketing for developing micro EVs, where range can be limited, thereby reducing the battery weight and where the vehicles can go round the corner for quick dose of charging. CTech has been working with IITB and other agencies like TIFAC (Composite Mission program) and various manufacturers to develop variety of appropriate applications for composite materials, particularly glass fibre-reinforced plastic, to effectively compete with traditional materials in cost, and offer superiority in function and performance. Environment & Public Design at CTech concerns itself with the design of public amenities, the design of public spaces, meeting points, hotel lobbies and plazas – making them user-friendly, amenity rich and artistically and culturally interesting iconic spaces through multimedia, digital sculptures, paintings.

## Patents:

https://blog.ipleaders.in/a-critical-analysis-of-design-protection-in-india/

## Commercialization

CTech licenses these products & technologies to interested companies for manufacture & marketing. Some of these are available for technology transfer to entrepreneurs looking out at new business opportunities and markets. https://www.ctechlab.com/ctechmart.html

# Keyless Secure Transportation System- Dr. L.N. Rajaram/Kirtilab Technologies, Chennai

Dr. L.N. Rajaram is Co-Founder, Kritilabs Pvt. Ltd. A graduate from IIT Kharagpur and PhD from BITS, Pilani, he was a pioneer in Deep Technology development for the past 35 years. He developed Complex Real Time Defence Software in the 1970s and 80s, Email engines in the late 80s and early 90s, Alerts Engine in late 90s, Mobile application Products in 2007 onwards. IOT platform-based Logistics Solutions from 2012 onwards and IOT,AI, ML Solutions from 2017 onwards.

https://www.linkedin.com/in/lnrajaram/

Srikrishna Varadharajan, MBA from ISB and patents owner is Co-Founder & CEO. KritiLabs received the Nasscom SME Inspire Awards 2023 for Leadership in Innovation - Tech Services.

## Technology

To help drivers more easily secure their trailers, Konexial has partnered with Nokē to create the My20 Locking System, a keyless, high-security, digital smart-locking platform used to provide trailer and container security for My20 ELD and Fleet Management customers. The My20 Locking System is secured by Nokē's smart-locking hardware and dual-layered encryption keys and is controlled from Konexial's My20 software ecosystem. The system can lock, unlock, track and control all lock access from anywhere, through the My20 iOS or Android mobile application. The My20 Locking system is available in options, the HD Padlock and Lock Block. The HD Padlock is made of hardened steel, has a 10 mm shackle and boasts a water-resistance rating of IP67. The lock body is designed to withstand tough environments and extreme weather conditions, and houses sophisticated technology. The Lock Block is a heavy-duty, steel housing that offers additional protection for the HD Padlock to provide an extreme-security, tamper-resistant, keyless locking system.

## Innovation

Intellilock was developed to tackle significant pilferage issues during transit in the oil and gas sector. Traditional locks were vulnerable to tampering, adulteration, and inefficiencies, leading to an approximate 2% loss per trip, amounting to substantial monetary losses. The founders saw a need for a solution that enhanced security and operational efficiency, providing real-time visibility and control over transportation assets. It is an OTP-Based Keyless Solution: Utilizes One-Time Passwords (OTP) for a secure and flexible unlocking system.

## Patent

Distributed Management System for security of remote assets, Patent number 10200858

Remote Asset Management System, Patent number 352900

Distributed management system for security of remote assets, Patent number 414377, Inventors: Rajaram Lalgudi Natarajan, Srikrishna Marur Varadharajan

## Commercialization

Kritilabs Technologies, founded in 2012 by Dr. L.N. Rajaram and Srikrishna Varadharajan, is at the forefront of the digital transformation revolution. With a strong foundation in IoT, AI, and Machine Learning, Kritilabs has been delivering innovative solutions to address inefficiencies and security challenges across various industries. The company's patented AI2oTTM platform is a testament to their commitment to providing cutting-edge technology solutions that redefine asset management and transportation security. Intellilock by Kritilabs Technologies has been implemented by various customers including Indian Oil, BPCL, Shell-MRPL, Hindustan Zinc, Sudha Diary, NAC Jewellers etc.

https://www.kritilabs.com/intellilock.html

# Micro Gasturbine-Rohit Grover/ Aerostrovilos Energy Private Limited/IITM

Rohit Grover with dual Degree (B.Tech+M.tech) Aerospace Engineering from IITM is Co-founder & CEO at Aerostrovilos. Pradeep Thangappan, graduate from Misrimal Nayajee Munoth Jain Engineering College is the co-founder. Prof. Satya Chakravarthy mentor to many startups, Founder CEO-CTO, The ePlane Co., Co-founder Advisor: Agnikul Cosmos, Aerostrovilos, X2Fuels, GalaxEye, TuTr Hyperloop, Advisor: Avishkar Hyperloop, NewTrace and Head, NCCRD. Prof. Satya received DRDO Academic Excellence Award, Dalmi-HEMSI-ACRHEM Award and HAL Prize.

https://www.linkedin.com/in/satya-chakravarthy-51326241/

National Centre for Combustion Research and Development (NCCRD) is founded by the Combustion Institute – Indian Section and funded by the Department of Science and Technology at Indian Institute of Technology Madras and at Indian Institute of Science, Bangalore. NCCRD underlines the growing recognition of combustion and its importance on energy and environment. The startup received first grant of around 1 crore from BPCL in 2018.

https://ioe.iitm.ac.in/project/microgrid-technologies/

## Technology

Micro Gas Turbine (MGTs) represent a kind of gas turbine with power levels up to 500 kw. A typical MGT consists of a compressor, a combustion chamber, and a turbine. Also, modern high-efficient MGTs are commonly equipped with recuperators to recover the waste heat. The airflow passes through the inlet, the compressor impeller, the cold passage of the recuperator, the combustion chamber, the turbine impeller, the hot passage of the recuperator and then exits. Based on this process, The ambient air is compressed into high-pressure air in the compressor and heated in the recuperator, where the waste heat of the exhaust gas is recovered. The air then enters the combustion chamber and burns, forming high-pressure gas at high temperatures.

## Innovation

The team developed its first gas turbine product, a 100kw micro gas turbine for power generation. It is a fuel flexible generator that can on practically any liquid and gaseous fuels. It is flexible to the extent of being fuel agnostic where the user doesn't have to bother about which is gaseous of liquid fuel line. The Lean Direct Injection (LDI) burner developed at NCCRD, IIT Madras, was utilized to convert the liquid bio-crudes derived from MSW and agro wastes via hydrothermal liquefaction to power. In the first phase, a 20 kw microgas turbine was demonstrated and the output characteristics tested with the microgrid. In the second phase, a 100 kw microgas turbine was developed for application at scale.

## Patents

IP developed at NCCRD for low emission and fuel agnostic combustion was transferred to the startup.

MICRO GAS TURBINE, WO2012175341A1, Inventors- BAUMGART THOMAS [DE]; LANGE ELMAR [DE]

Lean direct injection combustion system, CN101532679A, Inventors- BENJAMIN LACY; BALACHANDAR VARATHARAJAN; STEVE ZIMINSKY WILLY; KRAEMER GILBERT O; ALLEN BOARDMAN GREGORY; ERTAN YILMAZ; PATRICK MELTON

## Commercialization

100kw MGT, LX-101 is lightweight and compact and can fit in a 1cubic meter box. With only 1 moving part, the maintenance cost is low with annual maintenance check and minimum life of 10,000hrs or 10 year. https://www.aerostrovilos.com/product/

# Non-Destructive Measurement of Brix/ Concentration of Sugar Syrup-SAMEER

Tapas K Bhuiya, a scientist at SAMEER (Society for Applied Microwave Electronics Engineering & Research), has successfully led a team in designing and developing the "Microwave-Based In-situ Brix Measurement System" technology. Tapas Bhuiya holds an engineering degree from the University of Calcutta and an MTech from IIT Bombay. The technology has been transferred to the industry for mass manufacturing..

## Technology

In the sugar industry, the quality of sugar is heavily dependent on the concentration of sugar syrup during the boiling process. This concentration, known as Brix, measures the percentage of sucrose in the solution. Maintaining the right Brix level is crucial because if the concentration is too low, the crystallization rate is reduced, leading to lower quality and recovery of sugar. Conversely, a concentration that is too high can cause spontaneous nucleation, resulting in poor sugar quality. Traditional methods for monitoring Brix, such as electrical conductivity or refractive index measurements, have limitations in accuracy and speed.

## Innovation

SAMEER, with sponsorship from the Ministry of Electronics and Information Technology (MeitY), developed a microwave-based Brix meter to provide a more cost-effective and maintenance-free solution for the sugar industry. This Brix meter uses microwaves to measure the concentration of sugar syrup non-destructively and in real-time during the boiling process. The innovation includes a compact microwave transceiver system and an insertion probe designed to be mechanically robust and sustainable under the harsh conditions of a sugar mill. The system operates at 5.8 GHz and is programmable, allowing users to set transmission power and receiver gain through a graphical user interface (GUI). It has demonstrated an accuracy of ±0.3° in measuring Brix levels during trials in a live sugar factory

## Patents

**Paper:** *Instant In-situ Non-destructive Measurement of Brix/ Concentration of Sugar Syrup using Microwaves for Sugar Industry* This paper details the development and testing of the microwave-based Brix measurement system, highlighting its capability to provide instant and accurate readings in a live factory setting. The system's components include a microwave transceiver and a durable insertion probe that withstands the challenging environment of sugar mills.

https://ieeexplore.ieee.org/document/10464157

Patent database of MeitY:

https://www.meity.gov.in/writereaddata/files/patent-database.pdf

## Commercialization

The Microwave-Based In-situ Brix Measurement System technology has been successfully transferred to the industry. SAMEER, an R&D institute under MeitY, has signed a Transfer of Technology (ToT) agreement with Toshniwal Hyvac Pvt. Ltd. and Sir Automation Industries to facilitate the mass manufacturing of this innovative system.

# Quantum computing-Srinivasa Rao Aluri/ QNuLabs, Bengaluru

QNu Labs, a pioneer in the quantum cryptography field, was founded in 2016 by Sunil Gupta, Srinivasa Rao Aluri, Mark Mathias, and Anil Prabhakar. The company was incubated at the Indian Institute of Technology-Madras and later moved its operations to Bengaluru. Srinivasa Rao Aluri, a graduate in Pharmacy from BITS Pilani and an expert in business management with education from Osmania University, London School of Business, and Yale School of Management, serves as the Chairman of QNu Labs Pvt. Ltd. His extensive private equity experience includes turning around startups and identifying innovative investment opportunities. Sunil Gupta, a graduate from NIT Trichy, is the Co-Founder and CEO of QNu Labs. He previously served as COO at EdgeVerve (Infosys) and Paladion Networks. Prof. Anil Prabhakar, a faculty member in the Department of Electrical Engineering at IIT Madras and an expert in quantum science and technology, is also one of the co-founders. QNu Labs has a subsidiary, QNu Labs Inc, based in Massachusetts, USA.

In 2020, the Indian government announced a National Mission on Quantum Technologies & Applications (NM-QTA) with a budget of Rs. 8000 Crore over five years. This initiative aims to boost transformative technologies, including quantum computing, quantum communication, quantum key distribution, encryption, quantum devices, and more.

## Technology

Quantum cryptography, which operates at the single-quantum level, could be the first practical application of quantum mechanics. Unlike traditional cryptography, which relies on complex mathematics, quantum cryptography is based on the fundamental laws of physics. Quantum Key Distribution (QKD) is a cornerstone of quantum cryptography, enabling two parties to generate a shared secret key for secure encryption and decryption of messages. QKD uses quantum properties to detect any eavesdropping attempts, ensuring communication remains confidential and secure.

## Innovation

QNu Labs has made significant strides in quantum cryptography. After conducting successful field trials for its QKD product, QNu Labs received support from Cisco Launchpad in 2018. In 2020, the company launched two key products for both national and international markets: Armos, a quantum key distributor, and Tropos, a quantum random number generator.

## Patent

SYSTEM AND METHOD TO GENERATE RANDOM NUMBERS, Appl. No 201941002850, Applicant QuNu Labs Pvt. Ltd. Inventor - Anindita Banerjee

SYSTEM AND METHOD TO GENERATE RANDOM NUMBERS

## Commercialization

The startup introduced two products QRNG - Quantum Random Number Generator and QKD - Quantum Key Distribution. QNu's Armos QKD protects critical infrastructure unconditionally, providing quantum resilience to ensure data in transit is safe at all times. QNu' Tropos (QRNG) does not rely on mathematical algorithms; it generates random numbers from a quantum source, making it suitable for all applications.

https://www.qnulabs.com/qshield-platform/

# Rare-Earth Free Motor-Bhaktha Keshavachar/Chara technologies, Bengaluru

Chara Technologies, based in Bengaluru, is spearheaded by three industry veterans: Bhaktha Keshavachar, Ravi Prasad, and Mahalingam Koushik. Bhaktha Keshavachar, the Founder and CEO, is an Arizona State University graduate and former Intel veteran. He previously led motor design at Kirloskar Motors. Ravi Prasad, the Chief Motor Designer, brings over four decades of experience in building motors for various applications. Mahalingam Koushik, the Founder and CTO, is a graduate of IIT-Madras and Rensselaer Polytechnic Institute (RPI), NY, with deep expertise in power electronics, contributing significantly to Chara's innovative intellectual property (IP). The company is backed by investors such as Exfinity Venture Partners, Kalaari Capital, and IIMA Ventures.

## Technology

Chara Technologies focuses on developing rare-earth-free electric motors, specifically Switched Reluctance Motors (SRMs) and Synchronous Reluctance Motors (SynRMs). The SR motor, despite its simplicity and robustness, presents challenges due to its highly nonlinear flux-linkage

""

characteristics and torque production mechanism, making its analysis and controller design complex. Key limitations include the need for a power converter and rotor position sensing to start and run, as well as generating high torque ripple and audible noise. The Synchronous Reluctance Motor (SynRM) offers an alternative by using a rotor made of steel with air gaps cut into it to align with a rotating magnetic field. The rotor produces torque as it spins along with this field, with more torque generated when there is a significant difference in magnetism between materials (steel and air gaps). This design avoids the need for rare-earth magnets, which are often costly and environmentally challenging to produce.

## Innovation

Synchronous reluctance motors don't use magnets. Instead, a steel rotor with air gaps cut into it aligns itself with the rotating magnetic field. Reluctance, or the magnetism of a material, is key to this process. As the rotor spins along with the rotating magnetic field, torque is produced. More torque is produced when the saliency ratio, or difference in magnetism between materials (in this case, the steel and the non-magnetic air gaps), is greater. Chara is building a scalable, cloud-controlled, and rare-earth free platform for designing, building, and deploying electric motors for various applications.

## Patent

SWITCHED RELUCTANCE MOTOR WITH SEVERAL SINGLE-PHASE SLICES, Publication number: 20240154557, Inventors: BHAKTHA RAM KESHAVACHAR, RAVI PRASAD SHARMA, MAHALINGAM KOUSHIK BALASUBRAMANIAN

Abstract: The embodiments herein provide a multitude of single-phase Switch Reluctance Motor (SRM) arranged in a suitable fashion in order to achieve uniform torque with complete material utilization. The embodiment herein also provides a series of single-phase Switch Reluctance Motor (SRM) arranged in appropriate manner for better utilization of material leading to better efficiencies, reduced cost, reduced size or weight and reduction in torque ripple and noise. In addition, the embodiments herein also provide a two-slice SRM system and method, which correspondingly resolves the starting problem.

## Commercialization

With a DC bus voltage capacity of 48V to 72V, SYRGA Motor System delivers a peak power of 5.5kW to 8kW and a peak torque of 25Nm.

https://www.chara.co.in/solutions

# Retrofit for Diesel Generator-Kushagra Srivastava/Chakr Innovations, Delhi

Kushagra Srivastava graduate from IITD is CEO at Chakr Innovation. Echoing Green's Fellowship provided more than $2 million in seed. The startup has raised about $22 million (close to Rs162 crore) in funding.

## Technology

"Retrofit emission control device (RECD)" means Retrofit Emission Control Device (RECD) that has a particulate mass emission reduction efficiency, which qualifies it to be certified as meeting the classification class as defined in this Regulation. The regeneration system and strategy are part of the RECD. This includes any sensors, controller and software essential to the operation of the device. Systems that only modify the existing engine system controls are not considered to be RECDs.

Regulations- https://cpcb.nic.in/uploads/RECD_Procedure_Revised.pdf

## Innovation

Chakr had developed retro-fit emission control device (RECD) for diesel generators. The Chakr Sheild, a patented technology, reportedly captures over 70 percent particulate matter emissions from the exhaust of diesel generators. The retro-fit emission control device has received a type approval from the Central Pollution Control Board (CPCB). Their diesel fuel kit "enables a diesel engine to run on a mix of gas and diesel. It uses 70 percent gas and 30 percent diesel.

## Patent

PARTICULATE MATTER FILTER ASSEMBLY Patent Number: WO2022149168A1m Inventor: Konatham, Sunil Reddy Mishra, Avichal Pandey, Parth Sarthi Batham, Tushar R

## Commercialization

Built on catalyst-based positive filtration technology with no negative impact on DG set, the remarkable Chakr Shield efficiently prevents the release of harmful particulate matter into the atmosphere thus enhancing air quality and overall health.

**https://chakr.in/chakr-shield-product/**

# Ultrasonic Waveguide-Based Sensors-Dr. Nishanth Raja/XYMA Analytics/IITM

Dr. Nishanth Raja is the CEO of XYMA Analytics, a startup incubated at IIT Madras and a spin-out from the Centre for Non-destructive Evaluation (CNDE). Dr. Raja has an extensive background in ultrasonic waveguide technology, having worked as a Senior Research Fellow at the Fluid Control Research Institute and a project officer at CNDE, IIT Madras. He holds a Ph.D. from Anna University and has collaborated on several industry-related projects in ultrasonic waveguide-based temperature and flow measurements. The startup was co-founded by Aswin Kumar Kathirvel, a graduate engineer from IIT Madras. XYMA Analytics operates under the guidance of Prof. Krishnan Balasubramaniam and Prof. Prabhu Rajagopal from CNDE.

## Technology

Many industrial processes, such as crude oil refining and electricity generation, require operations at extremely high temperatures, often exceeding several hundred degrees Celsius. Traditional sensors like thermocouples and resistance temperature detectors (RTDs) face issues of accuracy due to sensor drift and mechanical failure over long-term use in harsh environments. For example, thermocouples provide localized temperature data and are prone to junction failure at elevated temperatures. In contrast, Ultrasonic Waveguide Sensors offer a more robust solution. They are constructed using the same material as thermocouple wires (e.g., Kanthal, Chromel, Stainless Steel) but without the vulnerable junction, reducing the risk of failure. These sensors have a smaller footprint and can be configured in various shapes—such as helical, spiral, or with multiple bends—allowing for distributed sensing over a wide temperature range

(30°C to 1400°C) and the simultaneous measurement of multiple parameters (e.g., level, temperature, and rheology).

## Innovation

The unique feature of XYMA's ultrasonic waveguide sensors is their adaptability and precision. They can be designed with different cross-sections (wire, rod, strip) and made from various materials (metals or ceramics) depending on the application. The sensors work by exciting an ultrasonic wave that travels along the thickness of the waveguide. Changes in the surrounding medium, such as temperature or humidity, affect the waveguide's material properties (e.g., thermal expansion, elastic modulus, and density), which in turn alters the wave's velocity, time-of-flight, and amplitude. By tracking these changes through the interaction of the ultrasonic waves with deliberately designed geometric discontinuities (like bends or notches), the sensors can provide localized information about the surrounding environment. This approach enables accurate measurements of temperature, viscosity, and other properties in challenging industrial conditions.

## Patents

IITM Technology Transfer Office, IPM Cell, IC&SR has signed a Technology Licensing Agreement Deal worth 2.5 Crore with an IITMIC incubated startup XYMA Analytics which is Licensed 8 patented technologies and one Trademark in the field of "Sensor-Based Technology for Process Efficiency In Industries".

Viscosity and Temperature Measurements at Very High Temperature by Ultrasound Reflection Viscosity and Temperature Measurements at Very High Temperature by Ultrasound Reflection, US 6,296,385

## Commercialization

**https://xyma.in/products**

https://www.linkedin.com/company/xyma-analytics/

https://xyma.in/

# Virtual Reality Based Technical Training- Srinivasan Yagnanarayanan/Square Comp Solutions, Chennai

Srinivasan Yagnanarayanan is Founder & CEO, Square Comp, has completed Seed Spark Program from Stanford Graduate School of Business and has over 13+ years of experience in various domains like Technology, Hardware, Marketing & Sales etc. Sriram Kesavan is Co-Founder & COO Sriram m an Automobile expert has completed his Diploma in Electrical & Electronic Engineering. He has over 10+ years of experience in Automotive sector under brands like Volkswagen, Mercedes Benz etc. The start up received support from Vel Tech-TBI, IIT Mandi and AIC-MUJ. They planned operations in USA- VR Headsets (the Hardware) will be manufactured in the Vel Tech TBI premises in Chennai and exported for use in US. The software functions, content creation & marketing for the American & European market will happen from Colorado.

## Technology

For Metaverse to become mainstream and play a role in our day to day activities, we need to have a mature technological ecosystem in it's core composition like Virtual Reality, Augmented Reality & Mixed Reality. All of these technologies in itself are in the beginning stages of their reign and will take some time to be adopted by the users for their day to day activities. The international device makers forgot to think about an estimated 64% of the Global Population who wear Glasses. These devices can't be used with the Glasses on and you will have to remove them if you want to experience VR. Whereas, our device can be used with the Glasses on which avoids the users from experiencing motion sickness even after prolonged usage of the device.

## Innovation

GRAHAs VR, a trailblazing brand under Square Comp Solutions (www.grahasvr.com), is at the forefront of transforming industrial training through immersive Extended Reality (XR) experiences. By harnessing the power of Artificial Intelligence (AI), Virtual Reality (VR), Augmented Reality (AR) & Mixed Reality (MR), GRAHAs VR is making training more accessible, affordable, and effective for businesses worldwide. Uniqueness of Zeal X: GRAHAs VR's Zeal X platform stands out due to its unique features and benefits:

➤ AI-Powered Content Creation: Empowers industries to create customized VR/AR training content tailored to specific needs and skill levels, reducing development time by 80%.

➤ Personalized Training: Offers personalized modules that adapt to each trainee's skill level, ensuring optimal learning outcomes.

➤ Enhanced Safety: Provides a risk-free virtual environment for trainees to practice and refine their skills without real-world consequences.

## Patent

The company has filed two patents for the device, one for utility and one for design vide Patent Application No. 202041045081.

## Commercialization

GRAHAs VR operates on a "Solution as a Service" model and the company is transitioning towards a platform- based subscription model to enhance scalability. With a growing clientele and with a strong industry presence, GRAHAs VR solutions are deployed across India & US, further expanding to Australia & Saudi Arabia. Some of the notable implementations include VOC Port, SIPCOT, Govt. of TN. etc. GRAHAs VR has received recognitions from leading Industry brands in India & Internationally like Startups Grind with Google, NASSCOM etc.

https://grahasvr.substack.com/p/grahas-vr-labs-xr-solutions-for-real

# Warheads-Satyanarayan Nuwal/Solar Industries, Nagpur

Satyanarayan Nuwal is a self made man, an entrepreneur par excellence. His story from DNA-Nuwal started renting explosive magazines in 1970 and started to make money from customers looking for ammunition to use in coal mines. Satyanaryan Nuwal eventually grew his company and became a consignment agent before rising to the position of greatest explosives dealer in India. Satyanaryan Nuwal finally gave this notion wings in 1995, conceiving of and founding Solar Industries as a result. Its headquarters were in Nagpur. Initially, Solar Industries supplied explosives to publicly owned coal mines. Afterwards, it made its own explosives and started working for the defence industry. The company has operations in 65 different countries and has its headquarters in Nagpur.

https://www.dnaindia.com/business/report-meet-satyanarayan-nuwal-from-sleeping-on-railway-platforms-to-establishing-rs-35800-crore-solar-industries-3038243

## Technology

A fragmentation warhead includes high explosive filling inside the fragmentation body. After the detonation of the explosive, high velocity fragments are dispersed from the warhead, and dispersed fragments may incur damage on the target although the target is not directly hit by the missile.

## Innovation

EEL, a subsidiary of India's Solar Group, is a prominent private-sector manufacturer specializing in explosives, Pinaka rockets, boosters, grenades, and various other munitions. The Nagastra-1 is derived from the Trinetra loitering munition developed by Z Motion Autonomous Systems, a startup in which Solar Group has a significant stake. Equipped with state-of-the-art GPS-enabled precision strike technology, the Nagastra-1 sets a new benchmark for accuracy in loitering munitions. Its striking capability, with an astonishing precision of 2 meters, makes it an ideal choice for engaging hostile targets effectively. The drone operates in two modes: the man-in-loop mode, which allows for real-time control within a range of 15 km, and the autonomous mode, extending its range to an impressive 30 km. This dual functionality ensures flexibility and adaptability in various mission scenarios. The Nagastra-1 boasts advanced day-night surveillance cameras, giving the operator superior situational awareness. This comprehensive visual coverage allows for accurate target detection and identification. Complementing its surveillance capabilities, the drone features a fragmenting warhead to neutralize soft-skin targets. This combination of surveillance and warhead technology equips the Indian Army with a potent and precise offensive capability.

## Patent

Paper- Innovative Technologies for Controlled Fragmentation Warheads,

https://asmedigitalcollection.asme.org/appliedmechanics/article-abstract/80/3/031704/370253/Innovative-Technologies-for-Controlled?

## Commercialization

The company currently produces everything from explosives and propellants to grenades, drones, and warheads as part of the Make In India mission. Developed SEBEX 2 utilizes a high-melting explosive (HMX) composition. This formulation significantly enhances the lethality of warheads, aerial bombs, artillery shells, and other munitions.

https://solargroup.com/DefenseProducts.html

# Warehouse Automation-Pramod Ghadge/ Unboxrobotics Labs, Pune

Pramod Ghadge, a graduate from VJTI with a postgraduate degree from the National University of Singapore, is the Co-Founder and CEO of Unbox Robotics. Before starting Unbox Robotics, he worked as a Robotics Manager at Flipkart. His co-founder, Shahid M., a graduate of GTU with a PG degree from the University of York, is the CTO. Shahid previously worked at Vanora Robots. Unbox Robotics was incubated at HAX, where startups receive at least $250,000 in initial investment and 180 days of close mentorship. The startup is backed by notable investors, including 3one4 Capital, Sixth Sense Ventures, SOSV, Entrepreneur First, Redstart Labs (Info Edge), Arali Ventures, CIIE, Beenext, WEH Ventures, Force Ventures, Veda VC, among others.

## Technology

Sorting robots are designed to automate the sorting process in warehouses and manufacturing facilities. These robots use a mix of sensors, cameras, actuators, and mechanical components to detect, sort, and place goods or parcels based on their destination, category, and other criteria. The sorting process is customized to meet specific business needs and is adaptable to different types of objects or parcels. Traditional automated sorting systems have included conveyor belts, Put-to-Light (PTL) technology, XY Gantry Robots, Articulated Arms, Mobile Robots, and others. Unbox Robotics leverages Swarm Intelligence—a concept credited to Gerardo Beni and Jing Wang in 1989 for their work on robotic systems. Swarm intelligence has since evolved into a field that integrates robotics, optimization, and distributed computing. Unbox Robotics applies these principles to develop decentralized, self-organizing robotic systems that solve complex problems in warehouse automation.

## Innovation

Unbox Robotics has developed autonomous robots using proprietary Swarm Intelligence and Machine Learning (ML) algorithms. These robots autonomously navigate through a warehouse or manufacturing facility, identify items that need sorting, and place them in their respective bins or locations. The robots are equipped with sensors and algorithms to determine the item's characteristics—such as type, size, shape, or weight—and make sorting decisions accordingly. The direct sorting mobile robots developed by Unbox Robotics are designed to execute sorting tasks based on a set of predetermined rules. Upon scanning an item, the robots use their rules to decide the item's appropriate destination.

## Patent

Pramod worked on the project 'Commercialisation of soft robots' at NUS and Commercialisation.

Promod and Shahid are inventors of the patent- System and method for automatically sorting items in a plurality of bins using robots, IN 201921014909 / EP3953069A2

## Commercialization

1st product of UnboxSort is the world's first of its kind patented AI-powered parcel sorting robotic system that gets installed in <2 weeks, saves ~50% - 80% area, and improves productivity by > 3X to 5X.

**https://www.unboxrobotics.com/solutions/**

# Wireless Semiconductor Systems-Chintu Singh/Ananant Systems, Bengaluru

Chitu (Chitranjan) Singh is Founder, CEO & President at LeapFrog Semiconductor and Founder of Ananant Systems. A product of IIT Kanpur and The University of Texas at Dallas, he started research at C-DoT, worked as Director of Engineering at Qualcomm before embarking on the entrepreneurial journey. Architected hardware and embedded software design and led the development of physical layer of industry's first 1Gbps wireless LTE Modem MDM9x55 used in the iPhone 8. Also, architected and led design of iterative demodulation feature supported in this modem.

## Technology

India's telecom networks depends on imported telecom gear accounting for nearly 90%. Development of indigenous technology got a major boost with government insisting BSNL to select local suppliers for 4g and 5G. BSNL ordered $1.83 billion worth of infrastructure equipment and software from IT company Tata Consultancy Services (TCS) for the deployment of its 4G network. The consortium of TCS (for Software support and System Integration), CDOT (for Core technology) and Tejas (for Radio network) was selected by BSNL for implementing BSNL's 4G project. TCS is the lead partner.

Indian army has inducted the first-ever indigenous chip-based 4G mobile base station, which it procured from Bangalore-based firm Signaltron. Weighing just 7 kgs, Sahyadri Network In a Box (NIB) systems provide high-quality secure wireless communication for audio, video and data application.

WiSig Networks, at IITH, is a pioneering supplier of advanced 5G O-RAN equipment, software, and services, our vision is to lead the industry with cutting-edge solutions. Offerings Include, O-CU and DU, O-RU; Advanced solutions, including massive MIMO and ULPI technologies with the highest radio capacity and coverage. Their Open Interfaces, enabling intelligent network operations for SMO and RIC functionalities. WiSig products are multi-vendor interoperable and ready for deployment in public and private networks.

## Innovation

Ananant Systems, a startup, is partnering with major telecom equipment makers to develop chips for BSNL's 5G network rollout. They aim to provide advanced connectivity, computing, and security solutions for 5G and 6G infrastructure. The company's innovative architecture promises to reduce costs and power consumption. They plan to focus on base stations in the first year and later move into device development. Ananant's partnership with IESA aims to establish India as a leader in wireless technology.

## Patent

CONFIGURABLE SUB-BAND FILTERING TO REDUCE PEAK-TO-AVERAGE POWER RATIO OF OFDM SIGNALS OR THE LIKE, Publication number: 20090323857, Inventors: Chitranjan K. Singh, Masoud Sajadieh

## Commercialization

Ananant Systems provides intellectual property (IP), chip design, semiconductor products, software, and systems – all developed and owned in India.

https://ananantsys.com/

# Profiled in 2022

## GROUP E- INDUSTRIAL PRODUCTS

1. Indian 5G technology- WiSig Networks

2. AUM (Air Unique Quality Monitoring)- CATS - CASTLE Advanced Technologies and Systems

3. Flash Memory- Sahasra Semiconductors (SSPL)

4. Plug and Play Mobile ESS- AmpereHour Energy

5. Waste Water treatment- Indrawater

6. Robot for cleaning Solar Panels- Enray Solutions

7. Autonomous Vertical Profiler (AVP)-NIO

8. Pothole Repair machine- CRRI

9. Solar PV Module- Sunora Solar

10. Roads with Plastic Waste- Prof. R Vasudevan

11. Bioleaching based Gold Extraction from E-Waste-Neshaju Envirotech

12. Veterinary Lifesaving temperature regulating machine- Thermodrip

13. Making existing Air Conditioners Energy Efficient- SMDPower Solutions (OPC)

14. Products from Carbon-Carbon craft

15. Worker Safety- Hacklab Solutions Pvt. Ltd

16. Facial Recognition App- MUKHAM

17. Touchless Technology-Hindonics Technologies

18. Smart Factory- Insightzz

19. Digital Manufacturing Platform-MEKR

20. Drones for Mine Management-Aarav Unmanned Systems

21. HPC Server- CDAC and VVDN

22. Nanofiber Solutions-E SPIN NANOTACH

23. Semiconductor Products- Saankhya Labs

24. Low Temperature Evaporation Technology-SPRAY ENGINEERING

## GROUP D- DEFENCE & AEROSPACE

1.  Anti Drone System -Zen Technologies

2.  Indian Global Navigation Satellite System (GNSS)- Accord Software & systems

3.  AR-Enabled Smart Helmet For Higher Security and Surveillance- AjnaTX Helmet

4.  Unmanned Tank - Big Bang Boom Solutions (BBBS)

5.  Green propulsion system- Manastu Space

6.  Bullet proof vest with graphene composite materials-Nanospan

7.  Electric Propulsion systems for low cost satellite launch vehicles- Bellatrix Aerospace

8.  Satellite launch: Agnikul Cosmos

9.  Thermal imaging-TONBO IMAGING

10. Smart Buoy- Saif Automations Services

11. High Altitude Pseudo Satellite (HAPS) UAV- NewSpace Research & Technologies (NRT)

# Profiled in 2023

## GROUP F: INDUSTRIAL PRODUCTS

1. AI based Video Analytics Platform-DocketRun, Hubballi,

2. Ajit Microprocessor- Prof. Madhav Desai, IITB

3. Automated Pencil Electrode Formation Pltform - Lanka Tata Rao, BITS Pilani, Hyderabad

4. Carbon Nanotubes- Nopo Nano Technologies, Bangalore

5. Carrier Ethernet Switch Routers- Prof. Ashwin Gumaste, IITB

6. Cellulose based geotextile - Anasua Guha Ray, BITS Pilani Hyderabad

7. Drop-In Liquid Sustainable Aviation and Automotive Fuel- Dr. Anil Kumar Sinha, CSIR-IIP

8. Electric Wheel Barrow-Technovos Machinery Pvt Ltd, Bangalore

9. Flow battery- Prof. Kothandaraman Ramanujam, IITM

10. GigaMesh - mmWave wireless backhaul radio-Astrome Technologies, USA/India

11. Guided waves sensors- Prof. Krishnan Balasubramanian, IITM

12. Horizontal boring machine for underground- CSIR-CBRI

13. Hot wire cutting machine- Prof. Sathyan S, IITM

14. Induction Motor Stethoscope (MSCOPE)- CSIR- CSIO and Ai-DEA LABS

15. Li-Fi products-Velmenni, New Delhi

16. Microfluidic Electro-Viscometer- Dr. Sanket Goel, BITS Pilani, Hyderabad

17. Silver Nanowire- CSIR-NCL

18. Ultra Spinner- CeraTattva, IITM, Chennai

19. Ternary Content Addressable Memory (TCAM) in a network router- Prof. Krishna Moorthy Sivalingam, IITM

20. Wirelsss GPS Clock- Signals & Systems (India) Private Limited (SANDS)

## GROUP E: DEFENCE & AEROSPACE

1. Aero Structure Components- LMW Advanced Technology Centre (ATC), Coimbatore

2. Air Independent Propulsion system- Naval Materials Research Laboratory (NMRL) and L&T

3. AI-based Aircraft Engine Inspection Tool- Awiros, Gurugram

4. BeagleZ-explosive detector - Prof. Anil Kumar, IITB/ NewTec Lab, Bangalore

5. Long Range Surveillance-Optimized Electrotech, Ahmedabad

6. Smallest short range VTOL quadcopter- Idea Forge, Mumabi

7. Satellite systems- Ananth Technologies, Hyderabad

8. Satellite Bus Technology, ISRO/ Alpha Design Technologies, Bangalore

9. Solar Ultraviolet Imaging Telescope (SUIT)- IUCAA,Pune

10. Smart policing- Staqu Technologies, Gurugram

11. Unmanned autonomous vessel- Sagar Defence Engineering, Pune

# Consumer Products

# Banana Leather-Jinali/Banofi, Kolkata

Jinali Mody leveraged her business experience from McKinsey &Co. and her Masters Degree in Sustainability from the Yale School of Environment to kick-start the venture. In her first year at the Yale School of the Environment, was looking for areas in climate and sustainability where a startup might be able to make a tangible difference. Where Ms. Jinali got $25,000 from Startup Yale's 2022 Sustainable Venture Prize and could make it a reality and purchase the necessary equipment for basic R& D and to build a functioning MVP. Banofi Leather won the $1 million Hult Prize in 2023. The Hult Prize is the world's largest student social entrepreneurship competition, over 10,000 teams applied for this year's challenge, Redesigning Fashion. Competition included spending 2 weeks at the Global Accelerator at Château de Jambville (a literal castle) and pitching at the historic Le Trianon theatre in Paris. The prize was awarded by the sustainable fashion legend, Stella McCartney.

## Technology

Piñatex is a leather alternative made from pineapple leaf fibres. This innovative material manufactured by Ananas Anam, featured in Vogue, is taking the fashion world by storm. The material is made by sun-drying long fibres derived from pineapple leaves, and then felting them together with a corn-based polylactic acid (PLA) to create Piñafelt—a base material. After that, special finishes are applied to the Piñafelt to achieve the range of commercial materials in Ananas Anam's collection, including Piñatex. Mirum is "created using NFW's patented plant-based curative – the first bioneutral alternative to petrochemical-based systems." Modern Meadow's Bio-Tex using Bio Alloy technology, which is produced through a unique fermentation process using natural materials like sugar and yeast. Desserto is made from the prickly pear cactus. Made from discarded skin and cores, Apple Skin looks similar to real leather but has a paper-like feel. Mylo is made from mycelium—the root structure of mushrooms, and Bolt Threads had even engineered a process to grow mycelium in a vertical farming facility powered by 100% renewable energy.

## Innovation

The material is made of banana pseudostem, polymers, adhesives, and dyes. The current composition is 65% banana fibre, 20% natural ingredients like rubber and gum arabic, and 15% synthetic ingredients like adhesives, acrylics and polyurethane. This has been through finding biobased replacements and systematically improving the pretreatment steps of the banana fibre. The majority of the material's properties, such as tear resistance (82.3N), flexibility (11.8 N/MM2), appearance, and feel, are comparable to those of animal leather. It has a thickness of 0.7-0.9 mm. The remaining 20 per cent is made from primarily recycled polymers required for the leather backing, but Atma is working to innovate and reduce its dependence on polymers. The leather has a significantly lower environmental impact than animal and plastic leather, with 90 per cent less carbon emitted and 90 per cent less water required in the manufacturing process. Atma's process also doesn't create any toxic wastewater and saves animals that would otherwise be slaughtered to produce leather.

## Patent

Application No. 202221047455,

## Commercialization

Product range includes- Coasters, Eyewear case, passport holder, penstand, key chain, card holder and bags.

https://banofileather.com/

# Insects as Food-Ankit Alok Bagaria/ Loopwarm, Bengaluru

Ankit Alok Bagaria with duel degree (BTech and MTech) from IIT Roorkee is Co-Founder & CEO at Loopworm. Abhi Gawri, graduate from IIt Roorkie is Co-Founder at Loopworm. ICAR-CIBA, Chennai signed a MoU with the team of M/s. Loopworm, for evaluating the utility of insect based protein and oil in aqua feeds. Funding support received from Titan Capital and angel investors including Nadir Godrej from Godrej Agrovet, Sanjiv Rangrass from ITC, and Akshay Singhal from Log9 Materials. ARSX is an internal competition to USDA Agricultural Research Service employees and awards $100,000 to winning teams with compelling research ideas. Among the three winners of this year's ARSX challenge seeking for bold, transformative ideas to "Harvest for a Healthier Future" is a project titled Harvesting Agriculture's "Natural" Insect Farms led by Alexandra Chaskopoulou, medical entomologist for the European Biological Control Laboratory (EBCL) in Thessaloniki, Greece.

## Technology

Protein and fat is the costliest nutrient in the aqua feed. With the rapid expansion of aquaculture, there is a growing demand for high-quality protein and oil in the aqua feed. Finding nutritionally appropriate and sustainable alternatives to fishmeal and fish oil is an applied research area that added many options to the aqua feed ingredient basket. Recently, insects have received significant attention as a new ingredient for aqua feed since they show many advantages such as low environmental impact, the ability to grow on by-products with better feed conversion efficiency, and a low risk of transmitting zoonotic infections. Insects are natural detritivores and a key part of the Ecosystem often ignored for their usefulness and considered pest & nuisance. They are a rich source of proteins, fats, and micronutrients, and an essential part of the natural diets of both land & aquatic fauna. They are an excellent source of high-quality Amino Acids, Fatty Acids, Peptides & Bio-molecules like Melanin & Chitin, having varied applications in Cosmetics, Nutraceuticals & Biopharmaceutical industries. The potential of insect use in animal and human food is mainly because there are more than 1 million known species. This generates innumerable possibilities and alternatives for its use; however, many studies are necessary since after identifying a species with potential, strategies should be created for production, reproduction, genetic evolution, and processing.

https://www.intechopen.com/chapters/70569

## Innovation

Loopworm is currently focused on producing alternative sustainable protein & fats for Shrimp Feeds, Poultry Feeds & Pet Foods. Highly digestible Amino Acids & insect-like natural smell in the protein concentrate lead to superior palatability in insect-based Aquaculture Feeds. Insect oil is a highly sustainable alternative to conventional feed oil sources like Palm Oil, Fish Oil,& Krill Oil. Inclusion of insect products in Poultry feeds leads to better Egg & Meat quality, higher productivity, better immunity, and enhanced welfare for commercially farmed birds. Hypoallergenic properties & natural bio-availability of anti-microbial supplements, cartilage & bone growth promoters lead to superior health for Dogs & Cats.

## Patent

Production and processing of insects for transformation into protein meal for fish and animal diets, US20080075818A1, Inventor- Ernest D. PapadoyianisXavier T. Cherch; Animal feed by applying black soldier fly larva, KR101878021B1,

## Commercialization

The startup successfully completed a major commercial consignment of Loop Meal and Loop oil totalling 30 Metric Tonnes with a leading shrimp feed manufacturer in India. https://loopworm.in/

# Low Glycemic Index Breads, Fortified with Dietary and Prebiotic Fiber-Amit Vaishnav/ Arihafoods Private Limited, Chennai

 Amit Vaishnav, is a graduate Mechanical Engineer from the Maharaja Sayaji Rao University, Baroda (Vadodara, Gujarat), batch of 1982.

## Technology

Grain free bread would be made from something such as almond flour. This kind of bread would be healthy for diabetes, and has started to get very popular among those that are going gluten free, or who prefer to eat no wheat. Spelt bread is similar to wheat. Sour dough bread is very tasty, and actually good for you. Acetic acid is produced during the fermenting process of this bread. Acetic acid is the main ingredient in vinegar, which helps lower blood sugar. This produces some good gut bacteria that aids in digestion, which then helps with regulation of blood sugars by slowing the rate at which carbohydrates are digested. Spelt bread is similar to wheat. This is one of the lowest glycemic index breads, and it will definitely give you more fiber, and keep your blood sugars steady over time. Pumpernickel bread is made from whole grain rye. Pumpernickel uses sourdough as a starter, and it's one of the healthiest breads for diabetes.

## Innovation

The Bread is formulated with ingredients that result in the final Glycemic Index (GI), which is indicative of how much and how fast the blood glucose level would rise / peak in the blood after consumption. The flours selected in the formulation have their GI within the mid-range of 50's GI. The formulation does not have any added sugars, starches, carbohydrates which can contribute to the total GI score. To improve the consumers "Gut Health" – the product formulation is fortified with addition of Crude Dietary Fiber as well as Prebiotic Fiber, this to the extent of providing almost 30% of the day's requirement of Dietary Fiber in one 100g Low GI Bread Loaf. While the baked Low GI Breads have the moisture in control in terms of the water activity – the product needs to be protected for preservation / longer shelf life. To enable this – the breads are packed using the Modified Atmosphere Packaging Technology and using a suitable packaging material that does not allow the ambient air / oxygen from getting into contact with the product to cause any microbial spoilage.

## Patent

liquid Sugar from refined cane sugar purifying process liquid Sugar from refined cane sugar purifying process. Patent number IN 201741040492 ·

## Commercialization

The 100g "Meal Substitute" Low GI Bread Loaf provides just about 200 kcal in a Low GI format as being fortified with 30% of the day's requirement of Dietary Fiber, the consumer does not experience any hunger pangs for up to 4 to 5 hours after consumption. In addition, the fortification of Prebiotic Fiber improves the Gut Health. Regular / daily consumption of the Low GI Bread Loaf as a Meal Substitute for minimum 1 meal per day will not only improve overall and gut health but will also help the consumer loose 2 – 3 kgs of weight per months during the initial months till the body stabilizes its dietary functions. The average increase of the Blood Glucose measured one and half after the consumption showed an increase of an average of 66 points as against an acceptable 120 units. This is tremendously beneficial for all diabetics and the calorie conscious persons.

www.amitvaishnav.in

# Play with Pixel-Rangarajan/Maphy Toys, Chennai

Rangarajan Srinivasan, MCA from University of Madras and MBA from IIM, Kozhikode is Founder & CEO of Maphy Toys, committed to enriching Children's playtime experience

## Technology

In today's digital age, where screens dominate childhood, we recognize the pressing need for toys that are open-ended, engaging, and foster exploration of each child's unique play style. Maphy aims to bridge this gap by providing toys that meet the evolving needs of children and their families.

## Innovation

"Explore the World" is a groundbreaking educational tool that transforms traditional geography learning into an engaging, hands-on experience. Designed to foster curiosity and enhance understanding, this product empowers both educators and learners to delve into geographical concepts in a dynamic and interactive manner. By combining tactile elements with advanced data visualization, "Explore the World" stands as a beacon of innovation in the realm of education. At the heart of "Explore the World" is the Geo Grid, an interactive platform that introduces the basics of latitude, longitude, position, direction, and distance. This grid forms the foundation for various educational activities, making abstract geographical concepts tangible and understandable. "Explore the World" features a variety of intuitive tools including data tags, basin tags, territory tags, quiz cards, route plotters, and monument icons. These elements enable users to visualize and interact with diverse geographical data, from river basins and population densities to significant historical monuments and country flags.

Play with Pixel An Open-ended platform to create and change pixel art with zero waste. A Starter-Kit, specially designed for beginners to get into the world of Pixel Art. A first step towards expanding your dream by adding more tiles and pixels. Download Play with Pixel app for pixelating your dream images, regenerating images based on your pixel color choices and to share and explore Pixel Art creations. This product is meticulously crafted from natural wood, with minimal use of additional materials.At Maphy, we prioritize environmental sustainability in both our products and packaging, aiming to minimize our carbon footprint.

Video: https://youtu.be/KepLtKzkZ_c?si=_wCm2ZcJ56KNVk40

## Patent

Pixel art crayon and marker photo array assembly and kit, US10118436B2, Inventor-Craig Skinner

## Commercialization

The product is designed for both educational and corporate environments. In classrooms, it can be used to create immersive geography lessons and interactive quizzes. In corporate settings, it facilitates team-building exercises and strategic data mapping, making it a versatile tool for various applications.

https://maphytoys.com/

# Translator for the Blind-Akshita Sachdev-Trestke Labs, Bengaluru

Trestle Labs' co-founder, Akshita, completed her degree in computer science in 2016 from Manav Rachna College of Engineering and started Trestle Labs, right out of college, along with her co-founder, Bonny Dave, empowering the blind and visually-impaired community towards inclusive education and employment by building solutions that enable them to Listen, Translate, Digitize & Audio'tize™ printed, handwritten and scanned-digital documents across 60 global languages. Recently awarded Residential Fellowship by Washington DC-based Halcyon Incubator, Akshita has received various National and International Awards including the DBS Foundation Social Enterprise Grant and Prosus Social Impact Challenge for Accessibility. Bonny Dave with degree in Mechanical Engineering from Nirma University is Co-founder, Trestle Labs. He worked on the project -Building an accessible environment for the visually impaired people- with Digital Impact Square, A TCS Foundation Initiative.

## Technology

Ministry of Social Justice & Empowerment established an Expert Committee to ensure that every child with a disability receives education in an "appropriate environment" until the age of 18. Trestle Labs is an Assistive Technology initiative, empowering the blind and the visually impaired community towards inclusive Education and Employment by building solutions that help them listen, translate and digitize any kind of printed, handwritten and digital content independently across 60+ languages.

## Innovation

Kibo, an acronym for Knowledge In a Box is a tool that allows users to Listen Translate, Digitize, Audiotize content. The core of the technology is converting scanned image based PDFs to searchable PDFs. The converted PDFs are Searchable and accessible even for Screen-readers used by visually-impaired users. A system to generate speech based on text forming part of a received image is disclosed. The system generates a clean image from the image received; detects number of columns in the clean image and stitches the number of columns in proper reading sequence to create a single column text image file carrying complete text of the image; extracts text from the single column text image file and performs text segmentation on extracted text to segment the extracted text into words, sentences and paragraphs to generate a segmented text file; and converts the segmented text file into a first speech output and provides the first speech output to a user according to the user commands. Content summarization, multimedia overlays and audio highlighting is provided.

## Patents

IN201921006144 - SYSTEM AND METHOD TO GENERATE SPEECH BASED ON TEXT FORMING PART OF A RECEIVED IMAGE, Application Number, Inventors- DAVE, Bonny, SACHDEVA, Akshita, BAGHEL, Abhishek, CHAKALASIYA, Jay Mohan, NAIK, Ninad Pradeep.

## Commercialization

Kuno XS kit-If you have hardcopy printed and handwritten documents and want to Listen, Translate, Digitize or Audio'tizeTM them.

Kibo 360 Kit- To scan hardcopy documents, upload images/PDFs or type text for real-time listening, translation, digitization and audio'tization.

Upto 90% subsidies for visually-impaired individuals through our partnership with VOSAP (Voice of Specially Abled People Inc.), a US based NGO that aims to make access to assistive technologies affordable for all.

https://www.trestlelabs.com/

# Solar for Home-Saurabh Marda/Freyr Energy, Hyderabad

Saurabh Marda, a graduate from RV College of Engineering with a postgraduate degree from GIT and an MBA from Yale, is the Co-Founder and Managing Director at Freyr Energy Services Pvt. Ltd. Prior to founding Freyr Energy, he advised CEOs of two large US solar companies in India, helping them secure over $30 million in orders with leading Indian solar developers. His co-founder, Radhika Choudary, is a graduate from Osmania University, with a postgraduate degree from Purdue and a fellowship from the Stanford Seed Transformation Program. She worked at Lanco Solar and SunEdison before starting Freyr Energy. Freyr Energy raised INR 58 crores in equity funding, led by EDFI ElectriFI, a European Union-funded impact investment facility, along with participation from Schneider Electric Energy Asia Fund, Lotus Capital, Maybright Ventures, and VT Capital.

## Technology

Freyr Energy operates in the domain of rooftop solar installations in India, which is regulated under the Electricity Act, 2003. Each state in India has its own net metering or rooftop solar policy that outlines the specifics of installing a grid-connected rooftop solar (RTS) system or small solar power plant. These policies dictate how consumers are compensated for the electricity their solar systems produce.

## Net Metering

Allows consumers to use the electricity generated by their RTS system and export any excess electricity to the grid. If consumers require more electricity than their RTS system produces, they can import the balance from the grid. **Gross Metering**: In this arrangement, all electricity generated by the solar system is fed into the grid, and consumers draw their electricity requirements from the grid. The Indian government, under its national rooftop scheme, provides a 40% subsidy for solar rooftop projects for households. Recently, Finance Minister Nirmala Sitharaman announced a new scheme to enable up to one crore households to receive 300 units of free electricity per month through rooftop solarization..

## Innovation

Freyr Energy has tackled several challenges in the Indian rooftop solar market, particularly in financing. In the early years, obtaining financing was a significant barrier due to the nascent state of the technology. Banks were hesitant, and non-banking financial companies (NBFCs) offered loans at high interest rates, making rooftop solar projects financially unfeasible. Since 2015, however, the rooftop solar sector has attracted over $2 billion in investments, mainly through equity funding (48%) and debt capital (29%). Nearly all equity investments (99%) have come from foreign entities eager to tap into the high growth potential and return on equity in the Indian market. Freyr Energy also benefits from international credit lines like the $625 million World Bank-State Bank of India (SBI) credit line and the Green Climate Fund (GCF)-Tata Cleantech credit lines, which specifically target India's rooftop solar sector.

## Patents

In USA many of the patents resulted from Government funding. In PV technology, in the period 1976-2018, this report identified a total of 63,172 patents (22,162 U.S. patents, 16,837 EPO patents and 24,173 WIPO patents).

## Commercialization

The company offers easy loans. A 3MW system costs around Rs2lakhs excluding subsidy. About 80,000 is available as subsidy. Loan is offered in 3 minutes with EMI payment spread over 60 months at zero interest.

**https://freyrenergy.com/solar-panels-for-home/**

# Strategy Games-Deepak Bharadwaj/ Lunchspace UNBOX, Gurugram

Deepak Bhardwaj, a seasoned deep-tech industry professional, has ventured into entrepreneurship to create educational games for children. With a background in mechanical engineering from BITS Mesra, Deepak began his career at Tata Motors Jamshedpur and later transitioned to roles at EDS PLM, Intel, Texas Instruments, and finally retired as Senior Vice President at Samsung. Notably, he was instrumental in setting up the world's largest mobile and India's first OLED mobile display plants, navigating complex technical, economic, and political challenges, including relocating plant and machinery from China to India and ensuring rapid construction and infrastructure development at the Noida facility.

Chandana Bhardwaj, Co-Founder of UNBOX, is an alumna of IIT Kharagpur with an integrated M.Sc. in Mathematics. She focuses on transforming middle and high school students' perceptions of mathematics through innovative games that incorporate mathematical concepts and fun elements. Besides her work in education and with UNBOX, Chandana is also an accomplished artist, known for her hyper-realistic portraits and works in various mediums, including acrylic, oil, charcoal, pencil, and digital art. Akhil Bhardwaj, another Co-Founder, is a graduate of Ashoka University and an investment professional.

## Technology

Card games are a powerful educational tool for children, even from a young age. Designed specifically for different age groups, these games support cognitive development by introducing basic concepts like numbers, colors, and shapes. Playing card games helps improve memory, concentration, and fine motor skills as children shuffle, deal, and handle cards. Additionally, these games promote social skills such as turn-taking, sharing, and following rules, fostering communication and cooperation among players. Moreover, card games often require strategic thinking, which helps develop problem-solving and decision-making skills.

## Innovation

UNBOX Z is Strategic Playing Card Game. Z is a strategic playing card game that redefines family fun and gameplay Perfect for birthdays, parties, or leisurely picnics with family and friends for players ages 10 and up, A new and unique card game that combines fast-paced action with educational value, Featuring a reimagined deck of cards, you can even make your own games with Z, With easy-to-understand rules, Z challenges players to strategize, group, and win, enhancing their number power while providing endless replayability.

Say goodbye to ordinary cards – Z is the ultimate reimagination of traditional card games, ensuring hours of laughter and excitement with friends and loved ones. Z is the culmination of numerous iterations and feedback from 9,000+ students, educators, and adults.

## Patents

Currently, the law does not qualify a card game that uses only a standard deck of cards for a patent. The game has to contain a new invention, such as a new deck of cards or a new game board alongside a new set of guiding rules, to qualify for patent protection.

## Commercialization:

Order- https://www.amazon.in/dp/B0D3RKVP31

Company webpage- https://launchspaceunbox.com/about-us

More- https://www.instagram.com/launchspaceunbox/

## Profiled in 2022

### GROUP F- CONSUMER PRODUCTS

1.  Double Fortified Salt- CSIR-CSCMRI

2.  Self-Heating technologies- Anchiale

3.  Home appliances powered by the BLDC motor.: Atomberg Technologies

4.  Super-Efficient ceiling fan- Super fan

5.  A mechanical Automatic Urinal-Toilet Flusher-NEERI

6.  Smart Intelligent Assistive Care Convertible Tech System - Coin Medix

7.  Paper Bag that can carry a weight upto 10kgs- La Fabrica craft

8.  Vehicle Engine Performance Monitoring-HiPER

9.  Solar Induction Cooker- Oxy Neuron India

10. Robotic-Assisted & Data-Driven Physiotherapy- Punar

11. Diabetic Monitoring with Smart Socks-FeetWings

12. Hygiene Management-microgo

13. Air Purifier- MedCuore Medical Solutions

14. zinc- based Battery (ZincGel)- Offgrid Labs

15. Germ destroying Air Filters- AiRTH

16. Audio Products- MiVi

## Profiled in 2023

### GROUP G: CONSUMER PRODUCTS

1. AC Helmet- Jarsh Innovations, Hyderabad

2. Board games- Mozaic Games, Bangalore

3. Burst Preventive Puncture Tyres - TJ Tyres, Faridabad

4. Mouseware- dextroware devices,Chennai

5. Piezoelectric microphone- Anand Richard Lobo, Goa

6. Robotic Scrubber Dryer- Aubotz Labs,Pune

7. Saline water lamp- Dr. PURNIMA JALIHAL, NIOT, Chennai

8. Siddu Jackfruit-ParameshaS.S and IIHR Karnataka

9. Stay Warm (Hand, Body, Foot and Sleeping Bag Warmers)- Parisodhana Technologies, Hyderabad

10. Walnut cracker- Mushtaq Ahmed Dar, Kashmir

# Young Innovators (2024)

# Goa State Innovation Council (GSINC)

## INTRODUCTION:

The Goa State Innovation Council is established by the Department of Science, Technology &; Waste Management, Government of Goa. Goa State Innovation Council engages in organizing various programs and events to spread the awareness about innovation and entrepreneurship among the people of Goa, handhold budding start-ups and innovators in scaling their business ideas and identifying potential ideas and innovations through events.

The Virtual Innovation Register is a unique initiative by the Goa State Innovation Council to harvest potential ideas and innovation in a very systematic manner. Keeping in line with the ethos of Digital India, the VIR is an online platform where innovators and entrepreneurs can register their ideas virtually and source the required support to achieve the expected results. VIR functions as an innovation bazaar where young innovators display prototypes and directly talk to prospective buyers. Virtual Innovation Register (VIR) was published in the E coffee table book on innovation titled 'Cutting-edge Transformations' launched by the Department of Administrative Reforms &; Public Grievances. The Hon'ble Prime Minister of India, Shri Narendra Modi, launched the E coffee table book at the Civil Services Day 2022 held on April 21, 2022, in New Delhi. The book recognizes VIR as a Digital Innovation Bazaar for innovators and entrepreneurs to register and secure their ideas online and be discovered by investors from around the world.

Why is VIR Good for Innovators & Businesses?

➢ Safeguarding unique innovations and ideas

➢ Validation of idea and support from experts

➢ Hassle-free digital registration from the comfort of home or office

Innovations and ideas can be registered under VIR in two categories; New Ideas and Startups. While the former allows individuals to submit their innovation and ideas, the latter allows already functioning start-ups to register with VIR and enjoy a host of benefits.

Benefits of Registering New Ideas Under VIR:

➢ Intellectual Property Rights support

➢ Support for commercialisation

➢ Pitching to prospective buyers

Benefits of Registering Your Start-up Under VIR:

➢ Collaboration with mentors and experts

➢ Support for raising Funds

➢ Access to resources (Incubation, Co-Founders, etc.)

Every Innovator or start-up begins with an Idea. And ideas are vulnerable. That's because there are more chances of them being killed early in its premature stage because they do seem absurd or unfamiliar. If they survive the ruthless process of evaluation, the chances are they won't get a proper platform that will help them to flourish. VIR is the platform where anyone can bounce their start-up ideas and get validation from experienced mentors who have significant industry experience. VIR also gives the Ideators easy access to sophisticated tools to evaluate the commercial viability of ideas.

**Rapid Prototyping Lab**

Equipped with the latest technology and tools, like an advanced 3D Printer and a powerful Laser Cutting Machine, the lab allows innovators to freely tinker around with ideas until they can refine them to the point of idealisation. To develop the spirit of innovation further, Goa State Innovation Council invites Students, Startups, Innovators, Research Faculty & Entrepreneurs to innovate, conceptualise, and scientifically shape their ideas. Our mission is to support prototyping and make it affordable for Students, Startups, Innovators, Research Faculty & Entrepreneurs who require the necessary support in converting Ideas into scalable products.

**The benefits Prototyping Lab are:**

1. Building the Product/Design Proofs

2. Saving Cost and Time

3. Customizing

4. Reducing Design Flaws

The Rapid Prototyping Lab is established at Don Bosco College of Engineering, Fatorda by the Goa State Innovation Council to provide innovators from all walks of life with the necessary infrastructure to transform their ideas into tangible models or prototypes. We provide access to the prototyping equipment to individuals with a purpose to convert the ideas into designs, and their designs into products.

**STATISTICS:**

- PROVISIONAL PATENTS GRANTED – 17

- PROTOTYPES BUILT – 57

- PROJECTS MENTORED – 88

- WORKSHOPS CONDUCTED – 583

- PARTICIPANTS TRAINED – 45,064

# Younf Innovators Supported By GSINC

## 1. Project Title: Smart glasses for Language translation

Group Members: Shivam Raikar, Vinayak Naik, Jesven Gomes, Vinayak Halshikar

Mentor: Ms. Vaibhavi Naik, Mrs. Manjusha Sanke

## Department: Information Technology

Institute: Shree Rayeshwar Institute of Engineering & Information Technology, Shiroda-Goa

## About the Project:

Language barriers ca use problems in our global world. People can't understand each other in education, business, travel, and everyday life. Translating languages can be hard, and conversations require translators or devices, making eye contact difficult. This project created real-time translation glasses to break down these barriers. The language translation glasses start with your voice input where it takes your voice as an input. Imagine you're talking with someone. The glasses have a microphone that captures your words, just like an attentive listener, and the microphone converts the user's voice into an electrical signal. Conversations received by the microphone aren't always crystal clear and hence a high-quality microphone is required, especially with background noise which thus increases voice/speech enhancement via Raspberry Pi 2w. A speech-to-text translation model is used for the conversion.

Further a model is used for text-to-text translation or language 1 to language 2. The model is trained on vast amounts of text in different languages. It swiftly analyses the meaning of the words and finds the match in the target language. The Smart glasses become a visual guide, projecting the translation onto their lenses, allowing you to read the translated text naturally and effortlessly as you interact with the world around you. These glasses could be very helpful in many situations, allowing people to understand each other during business meetings or casual conversations.

## Working Model

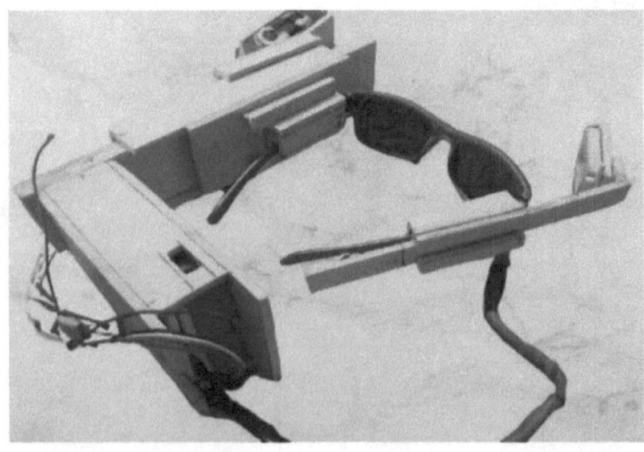

## 2. Project Title: "Tu Ulai"- Real-time Language Translation

Group Members: Mr. Anilgiri Goswami, Mr. Atharv Kindalkar, Ms. Samrudhi Naik

Mentor: Mrs. Pratiksha Shetgaonkar

Department: Computer Engineering,

Institute: Shree Rayeshwar Institute of Engineering & Information Technology, Shiroda-Goa

India is a hub of diverse cultures, with numerous native languages spoken across different corners of the country. As such, understanding the local language is a barrier to the communication process. In the context of India's linguistic diversity, effective communication across language barriers is vital. Konkani, spoken primarily in Goa, faces digital marginalization due to limited technological resources. This project, "Tu Ulai: Real-Time Language Translation," aims to bridge this gap by developing the first-ever real-time speech-to-speech translation system for English to Konakani language pair. Leveraging cutting-edge Natural Language Processing (NLP) and deep learning technologies, this system integrates advanced speech-to-text, machine translation, and text-to-speech components. The system converts a real-time translation of spoken words from English to Konkani, Marathi, and Hindi speech. By facilitating seamless communication between English and Konkani speakers, this project empowers Konkani speakers to access digital content and engage with global communities. This initiative promotes linguistic inclusivity and sets a precedent for developing language technologies for low-resourced languages, celebrating and preserving linguistic diversity in the digital age.

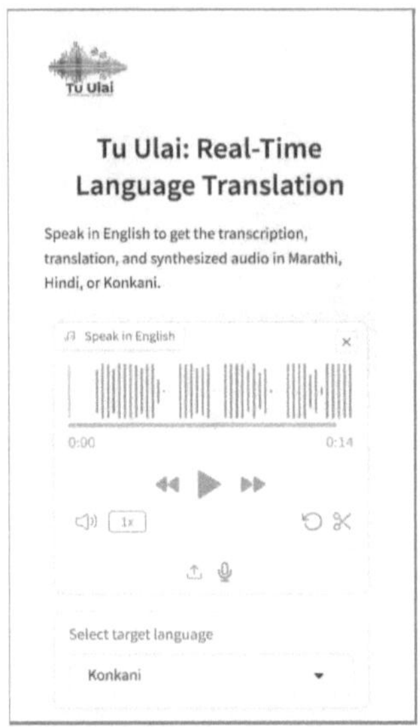


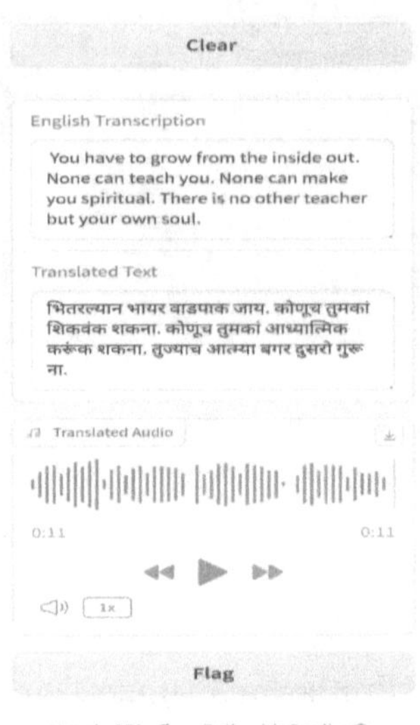


## 3. Optimization of scikit Library for Image Processing

Group Members: Virendrakumar Bind, Gururaj Mundargi, Tejash Naik, Vithoba Gawas

Project Guide: Chilton Fernandes

Department: Electronics and Telecommunication Engineering.

Institute: Shree Rayeshwar Institute of Engineering & Information Technology, Shiroda-Goa.

About the Project: The project titled "Optimization of scikit Library for Image Processing" aims to enhance the facial recognition capabilities of the widely-used scikit-image library. While scikit-image is known for its ease of use and minimal hardware requirements, it lacks robust facial recognition functionality. To address this gap, we developed a new library named myf-face-recognition.

This library integrates three advanced technologies: MTCNN for face detection in images, YOLO for face detection in videos, and FaceNet for facial recognition. The process begins with detecting faces in images or videos using MTCNN and YOLO. Subsequently, the detected faces are cropped from their backgrounds, processed, and fed into FaceNet for generating facial embeddings. These embeddings are stored for future comparison.

When a new image is introduced, the library detects and crops the face, generates the embeddings, and compares them with the previously stored embeddings. By setting a threshold, the system determines whether the faces match, thereby recognizing the individual. The myf-face-recognition library has been published on PyPI and GitHub, inviting further collaboration and improvement from the open-source community.

## 4. Development & Characterization of Green Composite from Cashew Nut Shell Oil.

Project Members -Amston Sanches, Ashish Tarale, Lemuel De Cunha, Nathan Mazarello, Chirag Naik

Project Guide- Prof. Gaurish M Samant

Composites have gained prime importance due to their high strength-to-weight ratios that make them applicable for various useful applications as compared to traditional metals. However, with certain advantages, these composites are a threat to the environment due to their toxic nature and non-biodegradability. Hence, a prime need of the hour is producing green composites made of naturally available fibres and bonding matrix materials. This project aims to produce green composites of three different natural fibres (Bamboo, Coconut leaf petiole and Areca leaf sheath), locally available in and around Goa, bonded together by Cardanol resin, a derivative of cashew nut shell oil. Silicon dioxide nanoparticles, graphene nanoparticles and single walled carbon nanotubes are added to increase its mechanical properties. Prior to composite manufacturing, the fibres are subjected to chemical treatments such as alkali treatment (NaOH) and Silane treatment to improve bonding properties between fibre and matrix, thereby increasing mechanical properties. A total of thirty-six composite samples are produced, where twelve are produced by Compression Moulding. This composite manufacturing technique utilises application of constant pressure onto the fibres laid in the matrix and results in the production of a composite of uniform thickness. Another set of twelve composites are manufactured by vacuum bagging technique which utilises vacuum pressure to cure resin-based composites by removing excess air and resin. The rest twelve by hand layup process, wherein fibres are placed in the matrix material by hand and kept for curing. Each sample is a combination of different fibres, treated or untreated, different nano-fillers and a matrix to hold them together. Each combination is obtained by the L9 Orthogonal Array (Four factors and three levels) of Taguchi's design of experiments. All samples are tested to obtain tensile and flexural properties by performing the tensile test and flexural test. Tensile properties such as Tensile Strength and Modulus of the whole composite can be obtained from the tensile test ASTM D3039. Similarly, Flexural properties such as the bending strength of the whole composite can be obtained from the Flexural test ASTM D790. The obtained properties are statistically analysed and compared with properties of already produced green composites. The composite mechanics is analysed in MATLAB and also, tensile and bending simulations are performed in ANSYS and COMSOL. Scanning Electron Microscopy (SEM) if performed to study the microstructure and morphology of the manufactured composites.

## 5. eNetraCare – One Stop Eye Care Solution

Name of Applicant/s: Shri. Abhay Bhamaikar

Name of Mentor: Atal Incubation Centre – Babasaheb Ambedkar Marathwada University, Aurangabad.

Name of School/College/Startup/Organisation: Innovease India Private Limited

Contact Number: 9923795140

Contact Email ID: abhay_bhamaikar@rediffmail.com

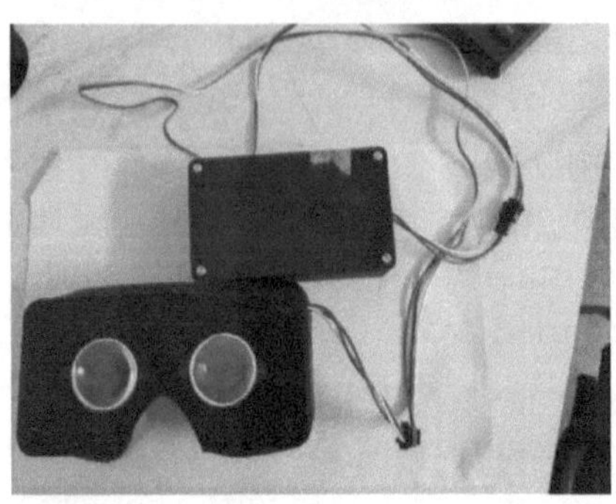

According to the National Programme for Control of Blindness survey, there is a backlog of over 22 million blind eyes in India, with approximately 80% attributed to cataracts. Despite being easily treatable, cataracts remain the leading cause of blindness and visual impairment in India, with an annual incidence of around 10 million cases. The lack of ophthalmologists and ophthalmic technicians in rural India exacerbates the issue, leaving many without access to proper eye care. In response, the promoter has proposed the development of a low-cost, AI-based mobile app cum screening device called "eNetraCare" to detect cataracts among patients aged 50 years and above, along with providing online eye care services.

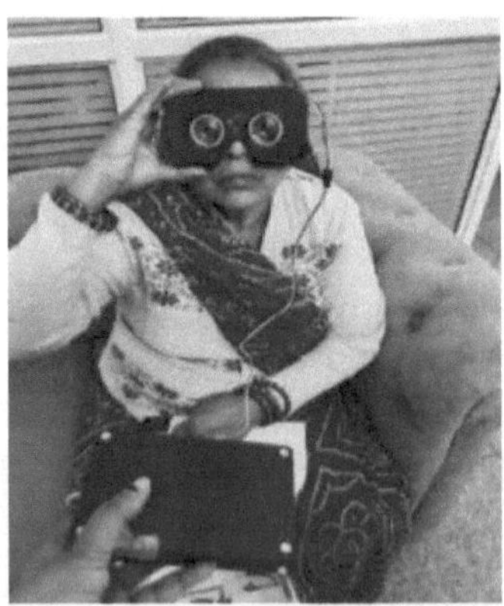

Project Outcome/result/findings: The project resulted in the development of a low-cost, portable, easy-to-use, 24X7 battery-operated vision and cataract screening device, along with an AI-based mobile app. Innovative Approach: The device is portable and battery-operated, ensuring accessibility and ease of use in various settings.

# 6. Wireless IoT Device for Hospitality

Innovator Name: Mrs. Supriya

Contact No: +919529962104

Contact Email: rachane.solutions@gmail.com

The Wireless IoT Device for Hospitality project aims to develop a wireless unit to facilitate better communication between customers and hospitality service providers, including restaurants, hotels, cafes, and resorts. The device leverages Internet of Things (IoT) technology to enable seamless and efficient interaction between customers and service staff, thereby enhancing the overall customer experience.

The wireless unit serves as a communication bridge, allowing customers to easily convey their preferences, requests, and feedback to the hospitality service provider. By incorporating IoT capabilities, the device enables real-time monitoring and management of customer interactions, leading to improved service delivery and customer satisfaction.

Key features of the Wireless IoT Device for Hospitality may include:

➢ Wireless connectivity: The device utilizes wireless communication protocols such as Wi-Fi or Bluetooth to establish connections with customer devices and service provider networks.

➢ Customer feedback system: Customers can use the device to provide feedback on their experience, rate services, and suggest improvements.

➢ Order management: The device facilitates the ordering process by enabling customers to place orders directly from their smartphones or tablets.

➢ Service alerts: Hospitality staff receive real-time alerts and notifications regarding customer requests, ensuring prompt response and service delivery.

➢ Data analytics: The device collects and analyzes customer data, allowing service providers to gain insights into customer preferences and behavior patterns.

Overall, the Wireless IoT Device for Hospitality project aims to revolutionize the hospitality industry by introducing innovative technology solutions to streamline customer interactions and enhance service quality. Through seamless communication and data-driven insights, the device contributes to creating memorable and personalized experiences for customers in the hospitality sector.

## Innovation Approach:

The development of the Wireless IoT Device for Hospitality involves a multidisciplinary approach that integrates cutting-edge technology with the specific needs and challenges of the hospitality sector. Leveraging expertise in wireless communication, Internet of Things (IoT), and user experience design, the project team focuses on creating a seamless and intuitive solution to enhance customer engagement and service delivery. Overall, the innovation approach adopted for the Wireless IoT Device for Hospitality emphasizes a holistic and customer-centric approach, aiming to revolutionize the way hospitality services are delivered and experienced. Through continuous innovation and collaboration, the project aims to drive positive outcomes for both customers and service providers in the hospitality sector.

# 7 Design and Development of UAV for harvesting Coconut

Project Members- Josten Dsouza, Samarth Savalkar, Clive Rodrigues, Delton Fernandes, Artika Vernekar

Project Guide: Prof. Gaurish M Samant

Precision agriculture has seen numerous advancements in recent times, including the use of unmanned aerial vehicles (UAVs) or drones. Drones are aircraft that operate without human pilots and can be remotely controlled or programmed to fly autonomously using embedded systems and onboard sensors such as GPS. In agriculture, drones have many applications, including precision farming, monitoring crop variability and even harvesting coconuts. Manual coconut harvesting is a dangerous and labour-intensive task, with a high risk of injury to the climbers. Furthermore, there is a shortage of skilled coconut tree climbers, making it challenging to harvest coconuts.

To address these challenges, a drone-mounted cutting mechanism was designed, including a harvesting mechanism, a quadcopter configuration frame, suitable landing gear, brushless direct current motors with propellers producing 28 kg thrust and a lithium-polymer battery with a capacity of 20000 mAh. A First-Person View (FPV) camera and transmitter are also installed to monitor the harvesting process, reducing the operation time. The quadcopter frame is made of carbon fibre, with a base plate and landing gear, allowing for precise harvesting of coconuts from high- raised trees.

The cutting mechanism system is generally attached below the base plate of the UAV, containing the grinder cutter or water jet harvesting mechanism. The grinder harvesting mechanism consists of a 18000rpm dc brushed motor, angle grinder blade mount, 1.8m aluminium square tube of 1x1 inch cross section, 1.5mm copper wire, 2200mAh 3 cell lipo battery 11.1Volts, relay module, FS-iA10B Radio receiver, abrasive metal cutting blade and a servo motor. The contact type harvesting mechanism cannot be taken too close to the coconut tree as there is a possibility of the propellers getting stuck in its leaves, hence we opted for the non-contact type harvesting mechanism.

## 8. Remote control vacuum cleaner and sanitizing machine

Name of Applicants: Viddesh S. Padiyar, Sharvari N. Chanekar, Saeesh G. Pangam, Vaibhav M. Naikoji

Name of Mentor: Prof. Suneeta Raykar

Name of College: Goa College of Engineering

The project is a wireless remote controlled moving vacuum cleaner and sanitizing machine with an ultrasonic sensor for obstacle detection.

➤ The overall machine can be battery powered or rechargeable. The sanitizer gets sprayed at regular intervals at given angle while the vacuum cleaner can suck in almost every considerable bit coming on its way (metal bits, dirt and non -metallic objects).

- The vacuum cleaner and sanitizing machine both components can be detached and used individually in different places.

- It can be used in any premises, majorly in hospitals, hotels. Even if not for this, it can be still used to clean and spread scent in rooms.

- It's a complete package within itself which is long lasting, user friendly and affordable in its segment.

## Project Outcome/result/findings:

The entire model specification in given below in the table

| Specifications | Value |
|---|---|
| Overall weight | 7.8 kg |
| Dimensions | 63.5cm x 40.64cm x 33cm |
| Vacuum cleaner weight | 2 kg |
| Sanitising machine weight | 2 kg |
| Remote controlled car weight | 3.5 kg |

## Vacuum cleaner efficiency

- Vacuum cleaner functions efficiently on full charged 12 v battery for 20-25 minutes.

- With increase in battery capacity the working time of vacuum cleaner can be increased. But overall weight should be considered while doing so. With increasing weight on car, the speed can reduce and over increase can damage the gear motor attached to tyres.

- De-attached vacuum cleaner can be used with similar ease.

- Also, with the use of DC-DC boost converter the suction power can be increased.

# Sanitising machine efficiency

- Sanitising machine functions efficiently on full charged 12 v battery for 30-35minutes.

- The distance covered by the sprayer is almost 30-50 cm.

- Due to the fault in power window motor, it does not attain the required rotating speed. It takes up to 9v as input which does not create enough pressure for spraying the liquid. Thus we were incompetent to find the spraying intervals and the actual speed of rotation.

## Innovative Approach:

- Considering the virus stays in air, it reduces the risk of infection to the cleaning staff in the contaminated areas (hospitals, covid centre).

- Addition of sanitising unit adds to the benefit

- It is detachable hence can be used individually

- Ultrasonic sensors prevent damage due to collision

- Easy to operate and transport and budget friendly

## 9. Utilisation of Ceramic waste in Concrete

Project members- MR. KAUSHIK S. FAL DESSAI, MR. RAVINDRA GURAV, MR. SAPNESH B. JANGLI, MR. ANVIT A. KELEKAR, MR. SAGAR S. LAWANDE.

Project guide- Prof. SATYESH KAKODKAR

The advancement of concrete technology can reduce the consumption of natural resources, which can be reused and find other alternatives. In India numbers of waste materials are produced by different manufacturing companies, thermal power plant, municipal solid wastes and other wastes. Solid as well as liquid waste management is one of the biggest problems of the whole world. Disposal of waste in to the land causes serious impact on environment. Nowadays large amount of tile powder is generated in tile industries with an impact on environment and humans. By using the replacement materials offers cost reduction, Energy saving sand few hazards in the environment. Concrete is nothing but a combination of aggregates both fine and coarse, Cement and water. Comparing to all other ingredients in concrete, cement is considered to be the expensive material. This is because cement is manufactured using energy-intensive process. Cement is one of the major producers of carbon dioxide, which is the main cause of global warming. During the manufacturing process of cement the formation of clinker can be achieved only by heating the cement at very high temperature. This leads to the release of enormous amounts of carbon in the atmosphere. This was one among the major problems identified for climatic changes. Various research works has been carried out for the cost reduction in construction with some of the locally available materials as the partial or full replacement material for cement. Over the last few decades supplementary materials like fly ash, rice husk, silica fume, eggshell, groundnut shell, etc. are used as a replacing material. These supplementary materials have proven to be successful in meeting the needs of the concrete in construction.

Ceramic waste is one of the most active research areas that encompass a number of disciplines including civil engineering and construction materials. Ceramic waste powder is settled by sedimentation and then dumped away which results in environmental pollution, in addition to forming dust in summer and threatening both agriculture and public health. Therefore, utilization of the ceramic waste powder in various sectors especially the construction would help to reduce the cost of concrete and also result in decrease of environmental burden due of recycling of ceramic waste. In this research study the cement will be partially replaced by ceramic waste powder accordingly in the range of 10% 20%, 30% by weight of M30 grade. M30 grade concrete will be prepared using 0.43 water cement ratio considering weight batching.

## INNOVATIVE APPROACH

Various tests were conducted to check the properties of the coarse aggregate and some of the tests including the specific gravity, fineness modulus, Initial & Final setting time etc. The Nanu crushed sand was used as the fine aggregate in this study. Various tests were conducted for fine aggregate also to find out the fineness of sand, specific gravity of sand etc. The test for cement was carried out to find the specific gravity, fineness etc. Mix Design of M30 grade concrete was carried out as per IS 10262-2019 and trials were conducted. Bharthi OPC 53 grade cement was used. Ceramic powder was used as partial cement replacement for making the concrete specimens. Asian paints admixture was used in concrete mix. Concrete Cubes were put for accelerated curing in an accelerated curing tank, 23 hours after casting. Concrete Cubes were kept immersed in a curing tank with water at 100 degree Celsius for 3 hours and later kept immersed in a cool water tank for 2 hours.

## 10. Railway Track Failure Detection System: Implementation Maintenance Strategies

Name of Applicant/s: Prithvi Amonkar, Rehan Khan, Saiprasad Parab, Raj Sawant, Mahesh Naik

Name of Mentor/s: Prof. Suraj R Marathe, Don Bosco College of Engineering Fatorda

Today' rail systems involve manual track inspection, which is cumbersome and not entirely effective. However, the detection and correction of track defects are a problem for all railway companies in the world. The objective of this project is to detect railroad track failures using ultrasonic sensor and GPS Module where once the crack is detected by the ultrasonic sensor the location is can be captured by the GPS Module Track maintenance follows traditional batter packing methods, but in most cases, trackers check for track defects by hitting the rod on the track judging the sound it makes. This method is completely based on experience and is not very accurate.

Project Outcome/result/findings: The final prototype model was placed on the track made of aluminium C channel for the purpose of testing the failure on the track. The coordinates of location where crack has been detected were sent to the web App and the location can be seen on map. Along with Location being shown on the map, the actual values of coordinates can also be Recorded from the web app.

Innovative Approach: The current detection system that is used by the railways is done manually. In our prototype model detection is done automatically with the help of ultrasonic sensor used for detection of failures. Also the advantage is after detection of the failure the exact location can be captured with the help of GPS module and the latitude and longitude and can be noted down which will be useful during the maintenance of tracks.

**Tactile Glove for Real-Time Bi-Directional Communication with the Deafblind, Chaitanya Tandon**

This innovative device is designed to facilitate real-time reciprocal communication between deafblind individuals and unimpaired persons. The glove utilizes capacitive touch sensors to detect touches and ERM vibration actuators to emulate touch and squeeze, allowing seamless bidirectional communication through a connected cellphone.The team comprises Chaitanya Tandon, a Computer Science Engineering major, and Nageshwar Kumar, an Electrical Engineering major, both from IIT Jammu. The invention won the Best Invention Award at the Invention Factory at IIT Gandhinagar, standing out among over 700 applicants. The program, run by the Maker Bhavan Foundation headquartered in Palo Alto, provided the platform to showcase and refine our innovative project.

## SkinCare AI: Personalized Cosmetic Suitability Analyzer

Project members- Gaddam Tejaswi, Meda Venkata Vasavi Swathi, Mirtipati Sravanth, Mohammed Abdul Sami

Project guide- Faculty: P Jyothi, Ravindra kumar

Institute- VNRVJIET, Hyderabad

The "SkinCare AI" project is a personalized cosmetic suitability analyzer that leverages machine learning techniques to predict the suitability of cosmetic products for various skin types. The system accepts input in the form of an image containing the list of ingredients of a cosmetic product and the user's skin type. Using image processing and optical character recognition (OCR) techniques, the system extracts the ingredients from the image. These ingredients, along with the user's skin type, are then fed into pre-trained machine learning models. The models are trained on a dataset of cosmetic products and their suitability scores for different skin types. Based on this input, the system predicts the suitability score of the given product for the specified skin type. The user receives the suitability score as the output, providing valuable insights into the compatibility of the cosmetic product with their skin type.

## Objectives:

1. Convenience: Empower users with a convenient and efficient solution to navigate the vast array of cosmetic products and find those that best suit their individual skincare needs.

2. Personalization: Provide users with personalized recommendations based on their skin type, enabling them to make informed decisions when selecting skincare products.

3. Compatibility: Develop a system to analyze cosmetic product ingredients and predict their compatibility with different skin types.

## Methodology:

1. Image Processing Module: It employs techniques such as optical character recognition (OCR) to extract the list of ingredients from the images.

2. Machine Learning Module: The machine learning module comprises pre-trained models that have been trained on a dataset of cosmetic products and their suitability scores for various skin type

3. Integration and Scalability: It may be integrated with existing e-commerce platforms, skincare apps, or beauty websites to provide seamless access to personalized skincare recommendations.

4. Recommendation Engine: Based on the extracted ingredients and the user's specified skin type, the recommendation engine calculates the suitability score of the cosmetic product.

## Conclusion:

"SkinCare AI" promises to revolutionize skincare by combining image processing, machine learning and personalized recommendations. While challenges such as data quality, image processing accuracy and user privacy must be addressed, the system's potential is immense. Future developments can enhance ingredient analysis and leverage advanced machine learning by expanding globally and collaborating with dermatologists, "SkinCare AI" can provide medically accurate, context-aware skincare advice. Embracing user feedback and integrating with e-commerce will further refine the system, making proposed skincare choices accessible and effective for users worldwide.

## ExoGear's monitoring solutions

ExoGear specializes in developing advanced monitoring systems designed to provide affordable and comprehensive technology solutions. Our focus is on creating versatile monitoring solutions for various parameters. Exogear also provides monitoring solutions for custom requirements.

Our products operate on Wi-Fi, enabling seamless data transmission to the cloud. This data can be accessed and managed through our intuitive web application, which is designed to be simple yet comprehensive. The application provides real-time analytics, generates PDF reports, and ensures that all essential information is readily available to users.

ExoGear's monitoring solutions are infrastructure-independent, allowing them to blend seamlessly into existing facilities. Our systems are scalable, supporting the installation of multiple units to meet the evolving needs of our clients. In addition to our monitoring products, ExoGear offers engineering consultancy services for customers who require customized solutions. Our expertise in the field allows us to develop tailored products that address specific requirements. We are proud to serve a diverse clientele, including Kreator, Vayut, Sytolabs, and Oceanatics. Our tagline, "Monitoring Simplified, Insights Amplified," reflects our commitment to providing streamlined monitoring solutions with enhanced insights.

https://www.exogeartech.com/

## AFFIX Empowering Students for a Brighter Future, SNIST

In today's fast-paced world, many students find themselves unaware of current trends and miss out on valuable opportunities simply because they lack access to a comprehensive network. This disconnect can have significant repercussions on their academic and professional growth. Enter Affix, an innovative and groundbreaking online platform designed to bridge this gap and transform the student experience.

Affix stands as a beacon of hope for students, offering a dynamic online application where they can thrive academically, professionally, and personally. This will also Enable students to build meaningful connections, access resources, and enhance their educational experience through a comprehensive digital platform. It aims to revolutionize the way students navigate the complexities of the modern world ensuring that no opportunity goes unnoticed or unexplored.

The platform fosters collaboration and networking, cultivating a vibrant sense of community among students. Through this innovation, students can learn from each other's experiences and perspectives, creating a rich tapestry of shared knowledge and support.

"Let's grow together as a community and build the connections that will shape our future,"

Contact:

AASRITHA KALLURI

(FOUNDER - AFFIX)

SREENIDHI INSTITUTE OF SCIENCE AND TECHNOLOGY

## PyroNEST, Karnataka

In contemporary times, significant advancements in science and technology have been observed globally. Modern India emphasizes these fields, recognizing their crucial role

in economic growth. India ranks among the top nations in scientific research, excels in space exploration with GSLV satellite launches, and has developed indigenous nuclear technology and ballistic missiles. Scientific knowledge drives technological innovation, solving practical problems and facilitating informed decision-making. Despite these advancements, considerable energy losses persist.

To address this, several projects focusing on energy conservation and alternative sources have been designed. "Current Gaining Lift" reduces electricity consumption in lifts by generating power through an AC dynamo, while "Current Gaining Engine" utilizes heat from vehicle engines to generate electricity via Peltier modules. "Generation of Power Using Road Humps" harnesses pressure energy from vehicles using piezoelectric sensors. Additional projects include "Charging Mobile Using Sound Waves," which converts sound energy into electrical pulses for mobile charging, and "Wireless Energy

Harvesting Using Rectenna," which transforms electromagnetic energy into direct current. The "Ethanol Petroleum Gas" project advocates for a high-performance, low- emission fuel blend of ethanol and petrol. These solutions aim to harness and conserve energy efficiently, contributing to sustainable technological progress.

LOHITHT.D, CEO & FOUNDER

Ph: 7892648995; lohithtd8@gmail.com; https://plyronest.com

## Publications

Books published by IIA:

1. Patent IPR Licensing - Technology Commercialisation – Innovation Marketing: Guide Book for Researchers, Innovators (2017),

2. Andhra Entrepreneurs: Past, Present and Future (2018)

3. Creating Demand for Local Innovations, Compendium of Best Practices (2019).

4. Top 100 Indian Innovations (2022)

5. Top 100 Indian Innovations (2023)

6. Top 100 Indian Innovations (2024)

   Details- http://motguru.com/, http://www.indiainvents.in/

   All books (hard copy and e-Book) are available on Amazon,

PATENT IPR LICENSING
TECHNOLOGY
COMMERCIALISATION
INNOVATION MARKETING
GUIDE BOOK FOR RESEARCHERS, INNOVATORS

Innovation
INDIAN INNOVATORS ASSOCIATION

COMPENDIUM OF BEST PRACTICES
CREATING
DEMAND
FOR LOCAL INNOVATIONS
INDIAN INNOVATORS
ASSOCIATION
CREATING DEMAND FOR LOCAL INNOVATIONS
INDIAN INNOVATORS ASSOCIATION

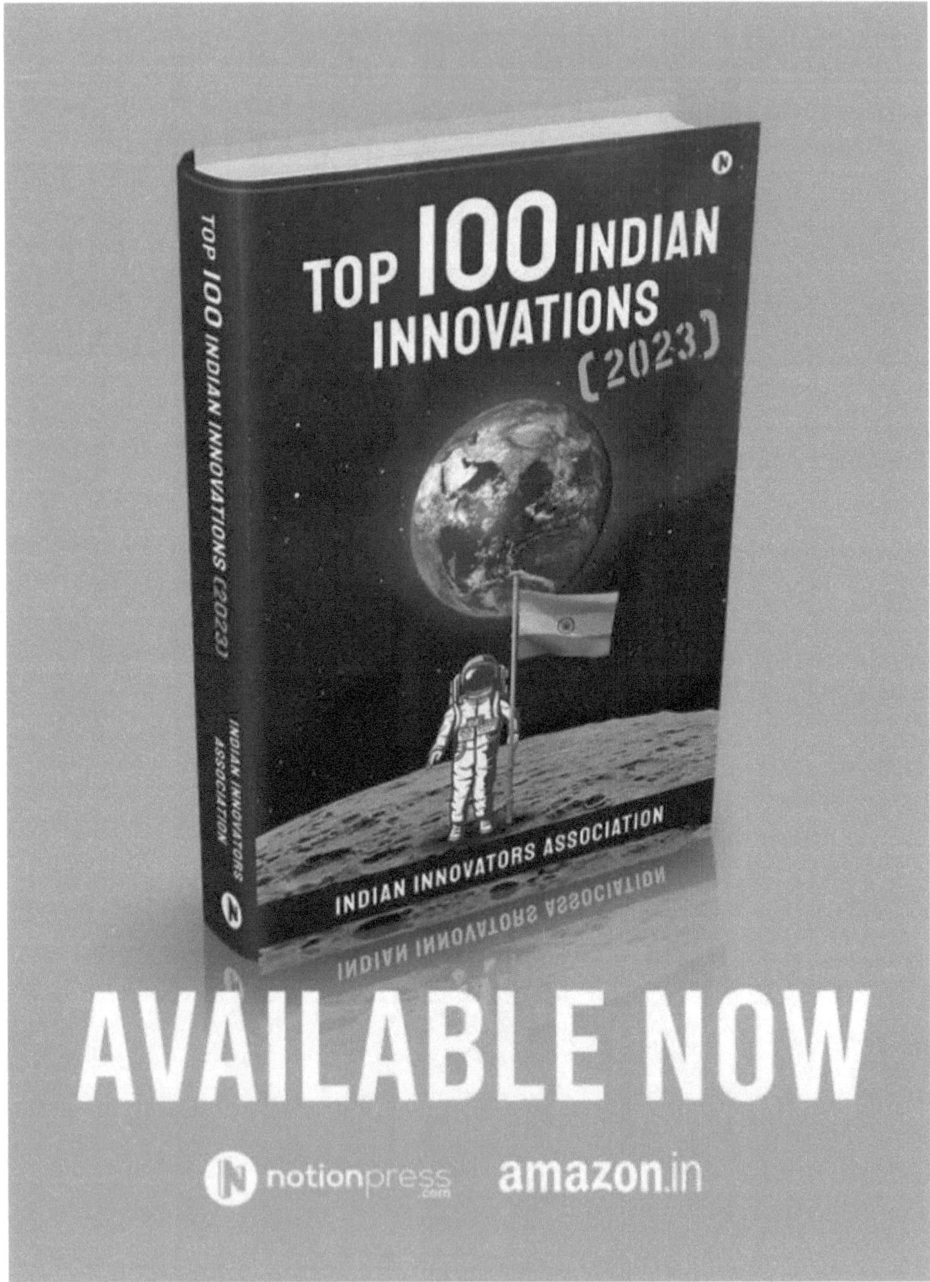
TOP 100 INDIAN INNOVATIONS (2023)
INDIAN INNOVATORS ASSOCIATION
TOP 100 INDIAN INNOVATIONS (2023)
INDIAN INNOVATORS ASSOCIATION
AVAILABLE NOW
notionpress.com
amazon.in

# TOP 100
# INDIAN
# INNOVATIONS
## (2022)

| INDIAN INNOVATORS ASSOCIATION |

## EXPERIENCE THE WORLD OF INNOVATIONS

- 30 Countries
- 300 + Inventions
- B2B Meets
- International Jury
- Awards
- Young E-NNOVATOR Pitch
- Seminars and conferences
- Hybrid Event
- HERITAGE Tour

### Honorary Partons & Partners for Erstwhile Editions

Ministry of Economic Development and Technology
Republic of Poland

Narodowe Centrum Badań i Rozwoju